NEC4: Practical Solutions

Preface

Following the resounding success of *NEC3: Practical Solutions*, the book *NEC4: Practical Solutions* was published to coincide with the launch of the NEC4 suite of contracts (June 2017). There has been a series of amendments made to the NEC4 suite of contracts since then, and this book has been revised to reflect amendments made in 2019, 2020 and 2023.

As well written as the NEC3 contracts were, there was a real appetite from users to obtain a better understanding of these contracts than others before, and therefore numerous questions as to their practical application prevailed. When analysed, it was evident that similar questions tended to be repeated – highlighting a potential gap in users' knowledge that might benefit from better guidance or more training.

NEC4: Practical Solutions provides a comprehensive set of responses to the questions that are most likely to be raised regarding the NEC4 suite of contracts. It is not, therefore, a clause-by-clause analysis, but addresses the issues, problems and queries that are expected to arise in use.

We have mainly based this book on real questions posed to the NEC Users' Group helpline over the past few years. We noticed that there were certain questions that occurred quite frequently, particularly concerning defined terms, the programme, payment, compensation events and Defined Cost. It is these recurring questions that have formed the basis for the book. Where necessary, we reworded NEC3 questions for the NEC4 contracts. *NEC4: Practical Solutions* provides detailed contractual responses to more than 280 common queries faced by practitioners during the design and implementation stages of a project, as well as the risks associated with running a project. As it mirrors the typical questions posed, there is a natural focus on the NEC4 Engineering and Construction Contract (ECC). However, the book also covers other contracts within the NEC4 suite.

The book is structured in the same chronological order as the ECC, which we hope will enable readers to use the book as a quick look-up tool to provide answers to common questions and problems that occur in practice. We have also included tips on better drafting of contract documents and made reference to further guidance to enhance its usefulness.

We wanted to write the book because, as NEC4 consultants, we recognise that the same repeat problems tend to emerge on projects. We are both passionate about providing concise and practical solutions to these common problems in a format that can be readily sourced. In turn, this should help project teams resolve differences in contract interpretation quickly.

It is our hope that *NEC4: Practical Solutions* will be an invaluable reference guide for engineers, surveyors, architects, project managers, supervisors, consultants, contractors, subcontractors and anyone else working with or interested in working successfully with the NEC4 suite of contracts.

About the authors

Robert Alan Gerrard BSc(Hons) LLM FRICS FCInstCES NECReg
Robert is a chartered quantity surveyor with 40 years' experience of setting up and managing contracts, mainly in the construction industry, in the UK and internationally. He is also an accredited mediator.

Robert has been the NEC Users' Group Secretary since 2005 and was part of the NEC4 drafting team. He is also an NEC consultant, and has run numerous NEC training courses, working with NEC contracts since 1996.

Patrick Waterhouse CEng FICE FCIArb FCInstCES MRICS
Patrick is a construction adjudicator, accredited mediator, expert determiner and contracts consultant. He is a renowned expert in NEC contracts, having worked with them since their first publication in the 1990s. He was part of the team that produced the NEC4 contracts. He has adjudicated and mediated many NEC disputes over the years. He is a member of ten adjudication panels.

Abbreviations

ALC	NEC4 Alliance Contract (June 2017, with amendments January 2023)
DBO	NEC4 Design Build and Operate Contract (June 2017, with amendments January 2023)
DRSC	NEC4 Dispute Resolution Service Contract (June 2017, with amendments January 2023)
ECC	NEC4 Engineering and Construction Contract (June 2017, with amendments January 2023)
ECS	NEC4 Engineering and Construction Subcontract (June 2017, with amendments January 2023)
ECSC	NEC4 Engineering and Construction Short Contract (June 2017, with amendments January 2023)
ECSS	NEC4 Engineering and Construction Short Subcontract (June 2017, with amendments January 2023)
FC	NEC4 Framework Contract (June 2017, with amendments January 2023)
FMC	NEC4 Facilities Management Contract (June 2021)
FMSC	NEC4 Facilities Management Short Contract (June 2021)
FMS	NEC4 Facilities Management Subcontract (June 2021)
FMSS	NEC4 Facilities Management Short Subcontract (June 2021)
PSC	NEC4 Professional Service Contract (June 2017, with amendments January 2023)
PSS	NEC4 Professional Service Subcontract (June 2017, with amendments January 2023)
PSSC	NEC4 Professional Service Short Contract (June 2017, with amendments January 2023)
SC	NEC4 Supply Contract (June 2017, with amendments January 2023)
SCC	Schedule of Cost Components
SSC	NEC4 Supply Short Contract (June 2017, with amendments January 2023)
SSCC	Short Schedule of Cost Components
TSC	NEC4 Term Service Contract (June 2017, with amendments January 2023)
TSS	NEC4 Term Service Subcontract (June 2017, with amendments January 2023)
TSSC	NEC4 Term Service Short Contract (June 2017, with amendments January 2023)

Robert Alan Gerrard and Patrick Waterhouse
ISBN 978-1-83708-959-8
https://doi.org/10.1108/978-1-83708-956-720252001

Chapter 1
General

1. Dealing with head office overheads in the Term Service Contract (TSC) (clauses 11.2(22) and 11.2(23))

QUESTION

How do we deal with head office overheads when using the TSC? Are these covered in the Fee?

ANSWER

There is no direct reference in the TSC to 'head office overheads', and this term might mean different things to different people. You need to understand what cost the *Contractor* can recover through the contract – this is called Defined Cost and is defined in clause 11.2(22) in Option A and 11.2(23) in Options E and F. Clause 52.1 then confirms that all costs not included in Defined Cost are treated as included in the Fee.

So, you need to concentrate on the components in the Defined Cost definitions and understand what they mean, and cover – such as people, Plant and Materials, Subcontractors and Equipment. Most of the components need to be Providing the Service within the Service Areas, but there are some exceptions. As you work through all other costs on the same basis, you will appreciate what cost falls within the definition of Defined Cost and that everything else is treated as included in the Fee.

2. Spirit of mutual trust and co-operation (clause 10.1)

QUESTION

Clause 10.1 sets out an obligation to follow what the contract states and, from clause 10.2, work in a spirit of mutual trust and co-operation. I appreciate that several cases have defined the latter.

Why are these separated into two clauses?

ANSWER

Yes, most of the contracts within the NEC4 suite contain the same clauses 10.1 and 10.2. These are separated in order to reinforce the need to undertake both.

- Clause 10.1 contains an express requirement to 'act as stated in the contract'.
- Clause 10.2 contains an express provision to act 'in a spirit of mutual trust and co-operation'.

Both aspects need to be complied with – not just the former. These provisions apply to all actions and procedures in the *conditions of contract*.

3. What is Site Information? (clause 11.2(18))

QUESTION

I have a query regarding the content of Site Information in an Engineering and Construction Contract (ECC) Option A contract used for design-and-build procurement. The project involves the design and construction of an asset, but the Site is to be determined by a developer and will be surrounded by a housing development and highways, which are yet to be designed in detail. Does the source of information impact its status as Site Information? For example, if a ground investigation or similar report is procured by the design-and-build *Contractor*, does it still constitute as being Site Information? Similarly, if a ground investigation report is provided by a third-party housing developer, could that be Site Information?

ANSWER

The first thing to say is that Site Information is a defined term (clause 11.2(18)), so only information that falls within that definition will count. Part of the definition refers to 'documents which the Contract Data states it is in', which means it must be referenced in the relevant entry in Contract Data part one.

So, to answer your question, the source of the information in the Site Information is irrelevant. If it is in the Site Information in the contract and complies with the first bullet of clause 11.2(18), then it is Site Information. The Site Information is, however, used in the ECC for only one purpose, which is to help determine what 'an experienced contractor would have judged at the Contract Date to have such a small chance of occurring that it would have been unreasonable to have allowed for them' (clauses 60.1(12), 60.2 and 60.3). Site Information is important in that it enables the *Contractor* to make reasonable assumptions at the tender stage, rather than assuming the worst and pricing for it. Clause 60.1(12) refers to compensation events that arise due to physical conditions at the Site.

You refer to a ground investigation procured by the *Contractor*. The status of that will depend on when it is procured and what the contract says. If the ground investigation is procured at the tender stage, and if the *Contractor* bases its price on the investigation, then it is sensible to include the investigation within the Site Information in the subsequent contract entered into. That gives better certainty to both Parties. In those circumstances, however, the *Client* should satisfy itself that the ground investigation has been competently carried out, as it potentially takes the risk for any physical conditions that are worse than shown. Of course, the benefit to the *Client* is that the tender price should be lower as a result.

Unless using secondary Option X22, there is no facility in the contract to change the Site Information after the Contract Date, unlike the Scope. Once the contract is formed, any ground investigation sourced by the *Contractor* does not become Site Information. The *Contractor* is, of course, entitled to argue that the information in its later ground investigation shows that the physical conditions are such as to trigger a compensation event under clause 60.1(12), but that will be judged partly on any Site Information you put into the contract.

Finally, NEC has produced some guidance on how to write the Scope for the ECC, which also includes a section on the difference between Site Information and Scope. This is contained in volume 4 of the ECC User Guides. We strongly recommend that anybody tasked with putting together both the Scope and Site Information reads this document.

4. What is the Early Warning Register? (clause 11.2(8))

QUESTION

I'm a new adopter of the ECC and note that there is an Early Warning Register. Can you explain the rationale behind this and how risks are managed in the contract?

ANSWER

The approach to risk management under the ECC is intended to be proactive rather than reactive. This is all contained within clause 15. The *Project Manager* is initially required to prepare a first Early Warning Register within 1 week of the *starting date*. This should be based on the identified matters from the Contract Data. The Contract Data also includes an entry to be made stating the frequency of early warning meetings in order that the matters are reviewed regularly. Many of the online communication systems available for NEC contracts include a semi-automated provision for the Early Warning Register, reducing the time commitment for the *Project Manager*.

Clause 15.1 places an obligation on both the *Project Manager* and *Contractor* to notify early warnings and potentially include others if agreed (clause 15.2). This may include the supply chain, the *Client*, stakeholders or both.

Clause 15.3 sets out how the matter should be managed (effectively forming the agenda for the meeting) and the need to co-operate and seek solutions that bring advantage to all those affected.

Clause 15.4 further requires the *Project Manager* to issue the Early Warning Register to the *Contractor* within 1 week of the early warning meeting and instruct a change to the Scope, if required at the same time.

This should ensure that the team focuses on proactive management and deals with potential problems effectively.

5. Establishing a cut-off between genuine cost in Providing the Service and what is effectively a head office cost in the TSC (clause 11.2(15))

QUESTION

I am working for a client about to procure an Option C maintenance contract over a large geographic area. There will be a lot of support work away from the Affected Property, such as the call centre, accounts and procurement.

How can we ensure that the *Contractor* does not charge for management and overhead costs that are included in the Fee?

ANSWER

Your aim should always be to ensure that the right amount is certified and paid; remember clause 10.2.

Clause 11.2(15) of the TSC defines the Service Areas, and this is relevant to assessing the amount due for some of the people used by the *Contractor*. There are five categories of people mentioned in the Schedule of Cost Components (SCC), which is relevant to main Option C.

- People who are employed by the *Contractor* and whose normal place of working is within the Service Areas. The costs of these people form part of Defined Cost wherever they are working. So, for example, if they attend a meeting somewhere else or visit a supplier then the costs would be allowable.
- People who are employed by the *Contractor* and whose normal place of working is outside the Service Areas – for example, people at head office or a regional hub. They are paid for according to the time spent in the Service Areas. Here, the management and overhead costs that concern you would not form part of Defined Cost.

- People who are employed by the *Contractor* and whose normal place of working is outside the Service Areas and who are working outside the Working Areas and who are identified in the Contract Data. For example, someone providing support on a technical helpline. These costs form part of Defined Cost and are paid at the rates stated in the Contract Data.
- People who are not employed by the *Contractor* but are paid for according to the time they work. Their costs form part of Defined Cost regardless of where they work. People in this category may be sole traders or those working through a service company but do not qualify as a Subcontractor.
- People who are employed by the *Contractor* and work outside the Service Areas in providing a *shared service* who are identified in the Contract Data as doing so. Their costs are paid at the rates stated in the Contract Data according to the time spent working.

Remember that for all costs listed in the SCC, they must be incurred in order to Provide the Service, so your scrutiny of the *Contractor*'s accounts and records should always ask why the cost was needed.

Item 41 (Subcontractors) within the SCC protects against subcontracting work that is covered by the Fee 'but not including any amounts paid to or retained from the Subcontractor by the *Contractor* which would result in the *Client* paying or retaining the amount twice'.

Finally, proposals to add to the Service Areas are dealt with in clause 16.4.

6. What is a Subcontractor? (clause 11.2(19))

QUESTION

We are using the ECC, and in relation to clause 11.2(19) we are trying to understand the cut-off between labour-only operatives and Subcontractors. Where is the line drawn?

ANSWER

What you must look at is the contract between the *Contractor* and Subcontractor. If the contract falls within one of the bullets stated in clause 11.2(19), then it is a Subcontractor, otherwise it is not. So, if it has a contract to lay all the brickwork on a people-only basis, for example, and is paid an amount per square metre to do so, then it is a Subcontractor under the first bullet. In that case, it has the obligation to provide the resources to do this within the time set out in the contract. If it instead has a contract to supply bricklayers at an amount per hour (and Equipment at an amount per hour), and those bricklayers will be put to work as determined by the *Contractor*, then it is not a Subcontractor because its obligation is to provide people to work rather than the result (last bullet of clause 11.2(19)).

7. Are in-house designers Subcontractors? (clause 11.2(19))

QUESTION

We have been on a fast-track ECC Option E contract, and the design was developed as the *works* proceeded. As a result, the designer had a heavy presence on Site as well as backup in the design office. The rates within Contract Data part two are now 5 years old, and the contract has an inflation provision for the designer. The designer is 'in house', albeit a separate legal entity under the umbrella of the parent company. There was no formal subcontract in place.

Within Contract Data part two there is a completed table for design employees (disciplines) under the line 'The rates for Defined Cost of design outside the Working Areas are'.

Under the ECC, should the designer be treated as a Subcontractor? If not, should the design works (done outside the Working Areas) be valued at the hourly rates in Contract Data part two plus the Fee plus inflation? Also, if not a Subcontractor, should the design works done inside the Working Areas be valued at the same hourly rates plus the Fee plus inflation?

ANSWER

This designer is not a Subcontractor because you have not got a subcontract with it (clause 11.2(19)). It is difficult to comment on the inflation provision, as no further details are provided. For design work carried out outside the Working Areas you will be paid for the designer's rate in Contract Data part two plus the Fee.

The designer's rates only apply for design carried out outside the Working Areas (item 7 of the SCC). If it is carrying out design in the Working Areas, your quoted rates are not used at all. Instead, you use the calculation for people set out in item 1 of the SCC, and to that you add the Fee.

8. How do we sign the contract? (clause 12)

QUESTION

We have a copy of the ECC, but there is nowhere for the Parties to sign. Is there another document that sits with this, or do we have to create our own signing provisions?

ANSWER

The omission is quite deliberate. It is for the Parties to decide how the contract is to be made, because it could well depend on the *law of the contract* (or even the law of the country that one or more of the Parties reside in). In England, it could also depend on whether you want a deed, or not.

So, you can make it as straightforward or as complicated as the applicable law requires. Offer and acceptance may be sufficient in the UK, or you can write out a more formal document (deed or not) for both Parties to sign. A suggested form of agreement is provided in Volume 2 of the ECC User Guides at chapter 6.

9. Mutual agreement to revise the Activity Schedule (clause 12.3)

QUESTION

We have an ECC Option A project on Site where the programme has been revised. As funding is time critical, we are questioning how we can ease the release of payments. Considering clause 32.1, can the *Project Manager* and *Contractor* mutually agree to revise the Activity Schedule to co-ordinate with the accepted revised programme, including splitting original contract activities within the Activity Schedule and tasks within the revised programme into two sub-activities (i.e. 'Delivery of materials on Site' and 'final complete installation')?

ANSWER

You can change the Activity Schedule because of a compensation event (clause 63.14), if the *Contractor* changes its planned method of working (clause 55.2) or if the *Contractor* corrects the Activity Schedule so that the activities on the Activity Schedule relate to the Scope (clause 55.2). None of these applies in this case. The only way that your suggestion can be achieved is if the *Client* and *Contractor* agree to change the contract and it is confirmed in the way set out in clause 12.3.

10. Instructing an additional *section* of the *works* (clause 12.3)

QUESTION

I am the *Project Manager* working on an ECC design-and-build contract. The *Client* has instructed me to notify the *Contractor* to accelerate a certain amount of work to ensure that it is completed on a certain date, which is before the current Completion Date.

Is the right way to do this to notify the *Contractor* of a new *sectional* Completion Date, providing the *Contractor* with a list of the work to be completed by this date and then ask the *Contractor* for a quotation for the impact of this?

ANSWER

You simply do not have the power under the contract to issue an instruction to complete *works* by a certain date or add a new sectional Completion Date to an existing contract. If you tried to do that, the *Contractor* could ignore your instruction with impunity, because it is only required to obey an instruction that is given in accordance with the contract (clause 27.3).

All that can happen is that the *Client* (not the *Project Manager*) and *Contractor* can agree to change the contract to accommodate the new requirements. It is likely that the *Contractor* will only agree to this for an additional sum, payment of which will form part of the agreement. Any such agreement must comply with clause 12.3. We therefore recommend that the *Client* needs to start talking to the *Contractor* about this, because the longer the agreement takes, the more expensive it is likely to be. As the *Client's* instruction applies only to a certain amount of the *works* rather than all of it, you cannot even use clause 36. And even if you could, you still need to get the *Contractor's* agreement to the quotation before proceeding.

11. Should we consolidate tender documentation? (clause 12.4)

QUESTION

We are about to start a contract using ECC Option B, which has been the subject of a prolonged period (6–10 months) of exchanges, refinements and discussions following the initial tender. We think it would be beneficial for a copy of a definitive document, which presents all the agreement after all the correspondence and negotiations, to be supplied before commencement. We consider this would make the contract management task simpler for both us and the *Project Manager*. We have asked for this but have been advised along the lines that we have all the information necessary already and, additionally, that it will be available by means of the electronic management system used for tendering. It would appear that the intention is to work with the multiple volumes of all exchanges/offers/counter-offers through the period as a package of emails rather than presenting the out-turn in a more concise package that references the necessary changes. Is there any written requirement in the ECC for a consolidation of such documentation?

ANSWER

We would recommend that you do not do this. For a start, there is simply nowhere for these records of your negotiations to be incorporated into an NEC4 contract. That is quite deliberate. And if they are not in the contract, then they probably will not be part of the contract (although that latter point is a legal matter that we cannot advise on). You need to start with a clear, unambiguous and consistent contract (as with all contracts), and you will not do it in the way suggested because the documents that form the negotiations will, inevitably, be inconsistent and often ambiguous, as Parties take certain lines and then subsequently concede them. Our recommendation is that once you have reached your agreement, you change the original documents (the Scope, Bill of Quantities etc.) to reflect that agreement, and use those as a basis for your contract. Case law (and, for that matter, adjudications and arbitrations) are littered with examples of cases where people have done as your *Client* has suggested, and the Parties have then, much later, fallen out over what they think the 'agreement' they made meant. This advice applies to all contracts under all forms of contract.

12. What is the status of verbal instructions? (clause 13.1)

QUESTION

We recently finished projects using the ECC in which we completed lots of additional work, but have not had instructions from the *Project Manager* (as verbally promised). The *Client* is refusing to pay any further monies until we show instruction and allocation sheets. What should we do?

ANSWER

Verbal instructions have no status in this contract – they must be in a form that can be read, copied and recorded (clause 13.1). You should not have acted on these instructions until they were in that form. And if the *Project Manager* has now proved to be unreliable by promising them in that form and then not delivering, you are potentially in trouble. As an aside, allocation sheets are not required for compensation events – you must make an assessment as to what the changes will cost, not keep records and use them afterwards (clause 63.1). We consider that you will need to take legal advice on furthering this matter. But as a clarification, the *Client* should pay what the *Project Manager* certifies. The issue here is with the *Project Manager*'s failure to instruct and certify.

13. Can you confirm verbal instructions? (clause 13.1)

QUESTION

We are using ECC Option B and have a query on confirmation of verbal instructions. We recall that confirmation of verbal instructions may be limited and has little or no contractual value and that every instruction must be given in writing by the *Project Manager* (or the *Supervisor* for those limited areas it is responsible for).

What is the status of confirmation of verbal instructions, and what are their limitations of contractual usefulness?

Clause 13.1 states that communications have to be in writing, so if a confirmation of verbal instruction was to be issued, and bearing in mind it is in a form that could be read, copied and recorded, would it be deemed as a contractually legitimate instruction if the *Project Manager* countersigned either simply with its name, or even stated 'Accepted' or 'Agreed'?

ANSWER

All communications that the contract requires are communicated in a form that can be read, copied and recorded (clause 13.1) or more probably through a communications system (13.2). In addition, notifications (of early warnings, compensation events etc.) must be communicated separately from any other communication (clause 13.7). All that means is that there is no such thing as a valid verbal instruction, and therefore no such thing as a confirmation of verbal instruction.

The *Project Manager* and the *Contractor* should understand that NEC4 contracts require you to communicate professionally. The *Project Manager* should not give an instruction unless it is willing to put it in a form that can be read, copied and recorded, and you should not accept or act on one unless it is in that format. If the *Project Manager* is not willing to do so, then it really should resign immediately, because it is not following the requirements of the *Client's* contract. That is the only safe advice we are able and willing to give you.

14. Should we act on verbal instructions? (clause 13.1)

QUESTION

We have an ECC construct-only project in which the *Contractor* built a car park. The design (included in the Scope) was provided by the *Client*. It turns out that there were problems with the design, and the *works* could not be built. Although written early warning notifications were raised, there were no instructions at the time from the *Project Manager* to change the Scope or any record of what was discussed at the early warning meetings. However, the *Contractor* acted on verbal instructions, and revised drawings from the *Project Manager*, which changed the Scope, and the *works* were built.

A year later, a compensation event was notified by the *Project Manager* as well as it formally changing the Scope to match the solution built. Therefore, there is no dispute about the change to the Scope being a compensation event, just a difference in how to prepare the quotation. In relation to the quotation, when should it be deemed assessed from (i.e. what part is on actual Defined Cost and what part is on forecast Defined Cost)?

ANSWER

Neither Party seems to have covered itself in glory on this contract! Verbal instructions are not valid in the ECC. The *Project Manager* should not have given them, and your colleagues should not have acted on them. It seems that the *Project Manager* issued drawings and so on for the new design, and that would probably be enough to say that the instruction was correctly issued at that time. When the verbal instruction was provided, the *Project Manager* should have issued the instruction correctly, should have notified this as a compensation event and instructed you to provide a quotation. This conclusion is reached reading clauses 14.3, 13.1, 61.1 and 61.2, in that order.

Therefore, the date of the verbal instruction and revised drawings issued is most likely the date you use to split 'actual' from 'forecast' Defined Cost, and to decide which Accepted Programme should be used. That means all of it is assessed using forecast Defined Cost. Quite deliberately, the dividing date (see clause 63.1) does not depend on when the assessment is made, because that will be different depending on who is doing the assessing: you, the *Project Manager* or the *Adjudicator*.

15. Using a communication system (clause 13.2)

QUESTION

We intend using a web-based system to manage communication on our ECC project. We find that using pro formas and registers can become complicated, and want to make use of modern technology. How do we encompass this in the contract? Is it a 'constraint' in the Scope?

ANSWER

We agree. High volumes of pro formas and contract communications can become difficult to manage. Clause 13.2 makes direct reference to the system you have identified within the Scope. It states: 'communication has effect when it is communicated through the system specified within the Scope'.

Volume 2 of the ECC User Guides (Preparing an ECC Contract) provides further advice on how to specify such a communication system.

16. Can the *Supervisor* be the *Project Manager*? (clause 14)

QUESTION

We are using ECC Option B. In this, the contract states that the *Project Manager* and the *Supervisor* should be distinct and separate; further, in another publication it states that there should be no line reporting from the *Supervisor* to the *Project Manager* and that the *Supervisor* carries out its specified duties independently of the *Project Manager*. It further states that even if there is line reporting between the two people named in their employment conditions, away from their ECC designated roles, this relationship cannot be mirrored in their duties carried out under the contract.

Bearing this in mind and acknowledging that the contract permits the *Project Manager* to delegate those duties it wishes to others, can the *Project Manager* delegate any of its duties to the *Supervisor*, or is there a conflict here with the intentions of how the ECC should be run, which would mean that these delegated responsibilities cannot be fulfilled by the *Supervisor*?

ANSWER

The roles and responsibilities are different. The skill sets are different. The *Project Manager* cannot just do in the ECC what the *Supervisor* does, and vice versa. The *Supervisor* is not subservient to the *Project Manager*, and the *Project Manager* cannot 'overrule' the *Supervisor* when it comes to the *Supervisor's* duties.

That said, there is no reason at all why the same person should not carry out the two roles on small/medium-sized projects. It is not at all unusual in those circumstances for the same person to be named as both the *Project Manager* and *Supervisor*. Equally, the *Project Manager* can delegate some of its actions to the *Supervisor* should it wish. There is no conflict with the intentions of those drafting the contract in either of these options. Having said that, if the *Project Manager* delegates all its actions to somebody else, then one questions whether that other person should have been named as the *Project Manager* in the first place!

17. Do we approve or accept? (clause 14.1)

QUESTION

Does it matter if we continue to use the term 'approved' when we sign off things such as the *Contractor's* drawings? This fits within our company quality assurance (QA) requirements.

ANSWER

It is important that you follow the requirements of the particular contract you are working with. In the ECC, for example, if there is design to be undertaken by the *Contractor* and submitted for acceptance by the *Project Manager*, then you are to 'accept' this, not 'approve' (clause 21.2). If you use words in your standard processes that are not associated with the particular contract, then have you discharged your obligation in that contract? This is further reinforced through clause 14.1, which states such acceptance does not change the *Contractor's* liability for its design. You may need to change your company QA requirements. You do, though, need to understand and follow the prescriptive requirements of each contract properly.

18. Could extensive *Project Manager's* instructions render our contract void? (clause 14.3)

QUESTION

We have an ECC contract with such a significant change to the Scope that we believe the contract to be essentially rendered void – what do you think?

ANSWER

Although you give very brief details here, we cannot offer legal advice so what we can say on this matter is in any case limited. The *Project Manager* has an apparently unfettered right to change the Scope (clause 14.3). However, the *Contractor's* obligation is to carry out the *works*, as defined in Contract Data part one (second bullet) (clauses 20.1 and 11.2(15)).

For example, if the *works* are described as to build a power station, it could be argued that the *Project Manager* could not instruct the *Contractor* to build an entirely and unconnected water treatment works, although it could instruct the *Contractor* to build one to treat the water coming into or out of that power station. In these cases, the *Contractor* is protected because the instruction will be a compensation event that is assessed based on a cost and time forecast. In addition, the Site is defined in clause 11.2(17) and the relevant entry in the Contract Data, and neither Party has the unilateral ability to change that definition. So, it could be argued that the *Project Manager* could not instruct anything to be built outside of that area.

There seems to be no doubt that the *Project Manager* can instruct extensive changes to the Scope, but how far it can go will end up being a legal question, which we cannot answer. And if it goes too far, whether that would render the whole contract void is also a legal question involving all sorts of complex legal principles.

19. What is the interaction between early warnings, the Early Warning Register and compensation events? (clause 15)

QUESTION

We would like to get clarity on the use and updating of the Early Warning Register within NEC4 contracts. We have an ECC contract using Option A, and the *Contractor* is claiming what we think is not an additional cost for the specified risk/compensation event.

Our contract stated that the *Contractor* can use some old apparatus for a new installation. This risk was recorded in Contract Data part one as a matter to be added to the Early Warning Register. Unfortunately, the old apparatus was damaged by fire and cannot be reused. A quotation was requested, and the *Contractor* is claiming things like Subcontractor transportation for its people to and from its Site and Working Areas. Our question is, how did the *Contractor* intend to do this work originally and what would be the extra in this case?

ANSWER

The first thing to point out is that the Early Warning Register and early warnings have got nothing directly to do with compensation events. They are about managing the risk of problems that may occur, whoever's they are. Compensation events are about events that have occurred or are going to occur and that are at the *Client's* risk. Not all early warnings will be about future compensation events, and not all compensation events will be preceded by an early warning.

We assume there was a *Project Manager's* instruction that changed the Scope, resulting in a compensation event arising under clause 60.1(1). The value of this compensation event is based on the forecast effect it is going to have on Defined Cost (clause 63.1). Therefore, you must decide how much the Defined Cost has increased. If this will mean installing new apparatus (you do not say), then the additional cost (if any) of the Subcontractor's additional people doing that might have to include any additional costs of getting them to the Site. That will be determined by inspection of the Subcontractor's contract to determine if such costs are recoverable (item 4 of the Short Schedule of Cost Components (SSCC)). It is difficult to be any more specific because we do not know what the transport arrangements are likely to be. However, if the Defined Cost of those travel arrangements increases because of this compensation event, that amount is payable by the Subcontractor to or for its employees and covered in the subcontract conditions, then the *Contractor* is entitled to recover that additional cost.

20. What do we do about the late notification of early warnings? (clause 15)

QUESTION

We have a *Contractor* on an ECC contract that is notifying early warnings well after the event. We believe that it is doing this in a cynical attempt to overload the *Client's* management team and achieve a tactical advantage. It probably believes this will end up in adjudication. We will be able to demonstrate that it has not attempted to minimise our costs at all over the duration of this contract. Can we as the *Project Manager* reject an early warning, and what benefit is there for any Party to notify early warnings after the event?

ANSWER

Clause 15.1 is clear, and it appears to be misunderstood here. You only give an early warning of a matter that 'could' have the listed effect. If it has already had the effect, then it is not an early warning; instead, it is a late warning! So, the timing depends on when the effect occurred, not when the event did. You must remember that the notifying of early warnings is not a one-way street. The onus is as much on you, the *Project Manager*, to do so if you are aware of any matter too, and if you do not, you will be in breach.

Early warnings are a form of risk management. They are not compensation events and do not entitle the *Contractor* to any more time or money. Notifying an early warning is not the same as notifying a compensation event. The *Contractor* still must do the latter, even if it has already done the former; and it has to do it within the 8 weeks set out in clause 61.3. The only proviso to that is that none of that applies to compensation events that result from your instructions, when the contractual onus is on you to notify (clause 61.1) and the 8-week time limit does not apply.

Having said all that, our view is that it is always in the *Client's* interests if the *Project Manager* is seen to be actively considering and discussing all early warnings, whether they are strictly contractually valid or not.

21. How do we deal with the lack of early warning notifications? (clause 15)

QUESTION

We are using Engineering and Construction Subcontract (ECS) Option A. There is a provision within the contract that enables the *Contractor* to consider early warning notices (or more particularly, the lack of) when assessing compensation events using clauses 61.5 and 63.7. Assume a Subcontractor risk event occurs, and an early warning is notified as it could affect the programme – clause 15.3 requires the Parties to co-operate to ensure, as far as possible, that actions are taken, and decisions are made that avoid or mitigate the effects of identified risks on cost, quality and time. In this case, the Subcontractor requires assistance from the *Contractor* in the induction of operatives and acceptance of method statement changes. The *Contractor*, unfortunately, is not forthcoming with assistance, and instead waits for the Subcontractor to sort the issue, but the effects of the event could have been reduced if the assistance from the *Contractor* had been received.

With reference to the above timeline and the provision in clause 63.7, is it conceivable that the delays are additional to those that would have been incurred anyway as a result of the risk event occurring? For example, a risk event occurs, *Contractor* assistance is received, total delay = 1 day. Or risk event occurs, *Contractor* assistance is not received, total delay = 3 days.

In the above example, is it possible to pass the 2 days' additional delay back to the *Contractor* – as had the *Contractor* acted as stated in the contract, this would have mitigated the delay? We are aware that there is no express clause in the contract for this, it is just that the contract will punish the Subcontractor for not notifying an early warning but does not punish the *Contractor* for not providing as much assistance as is reasonably practicable.

ANSWER

The main way that the Completion Date or Prices change is if a compensation event occurs. And we cannot see how one has occurred. And there is not one that says, 'the *Contractor* has not helped me to sort my problem out'. We assume that you were aware of the need for these inductions and acceptances – either they were in the Subcontract Scope, or they were part of the law. In that case, you were required to sort them out and they were your obligation.

Also, remember that the early warning and compensation processes are entirely different. The clauses (61.5 and 63.7) that you refer to are about you not notifying an early warning of something that was at the *Contractor's* risk, in the sense that it would turn into a compensation event. This is something that is entirely at your risk and that you knew about but did not manage properly. We know that sounds harsh but, as we say, we assume that you knew about it and yet you still let it delay the work, in the sense that you did not allow enough time for it. So, for the reason in our first paragraph, we cannot see how this can be a compensation event.

22. Value engineering (clause 16)

QUESTION

We are a *Client*, about to let a major contract under ECC Option A. Despite our best endeavours to draft the Scope, we consider that there could be significant value engineering savings both in terms of capital cost and whole-life cost. However, our procurement process has precluded early contractor involvement.

What provisions exist in the ECC to help generate such value engineering ideas post-contract?

ANSWER

The ECC deals with this in two areas. Firstly, clause 16 allows the *Contractor* to propose changes to the *Client's* Scope in order to reduce the total of the Prices. Within 4 weeks (clause 16.2) the *Project Manager* decides (having consulted with the *Client*). This effectively deals with reductions in capital cost.

The whole-life cost aspect is dealt with in secondary Option X21. Clause X21.1 allows the *Contractor* to propose reductions in the cost of operating and maintaining an asset over its life. The details contained within the quotation are detailed in clause X21.2 (i.e. a detailed description and analysis of the resulting risk). The *Project Manager* then replies within the *period for reply* with either an acceptance or reasons for not accepting (clause X21.3). For example, this could include a proposal for a more expensive pump in a pumping station that has a higher capital cost but a lower whole-life cost.

In summary, capital cost value engineering proposals are dealt with in clause 16 and whole-life proposals in Option X21.

23. How do we deal with ambiguities in the Engineering and Construction Short Contract (ECSC)? (clause 17)

QUESTION

Clause 17.1 of the ECC deals with ambiguities and inconsistencies. Can you please advise how ambiguities are dealt with under the ECSC?

ANSWER

The ECSC is designed for simple low-risk work, and therefore there should hopefully be fewer ambiguities. The only place this is likely to be is in the Scope. In that case, if it is unclear or uncertain, you should ask the *Client* to clarify what is meant. Any instruction given will be a change to the Scope, which would be a compensation event (clause 60.1(1)). In assessing that compensation event, it will be assumed that you used in your tender the interpretation most favourable to you (clause 63.10) – that is, the one that cost the least.

To give a practical example, if one part of the Scope says you use a particular item that costs £100 and another specifies, for the same thing, a different item that costs £200, you will be assumed to have allowed for the cheaper one. If the *Client* says, 'I want the expensive one', then the value of the compensation will be £100 plus the stated percentage uplift (assuming all other Defined Cost is the same).

24. Can a staff agency be called a Subcontractor? (clause 11.2(19))

QUESTION

In the ECC, if the *Contractor* procures staff from an agency, can the agency be classified as a Subcontractor?

ANSWER

To be a Subcontractor, the person or organisation must meet one of the three bullets in clause 11.2(19). All the agency will likely have here is a contract to provide people at the direction of the *Contractor*. It could be argued that the agency is providing a service in the Working Areas, but this third bullet specifically excludes the supply of people paid for by the *Contractor* according to the time they work.

25. Omitting work (clause 14.3)

QUESTION

Can we omit work from the Scope? I've been advised we need a specific clause stating as much.

ANSWER

The *Project Manager* has the power to give an instruction that changes the Scope. There is no restriction on whether this can be an addition, an omission or an alteration. So yes, work may be omitted but general law may step in to restrict certain additions, omissions or alterations. As always, it depends and we cannot give legal advice.

FURTHER THOUGHTS AND TOP TIPS

- Make sure when drafting the ECC Scope (and the equivalent in other NEC4 contracts) that you use the same terminology as laid down in the contract (i.e. 'Defined Cost' and 'Accepted Programme').
- The *Client* needs to think about appointing a suitable *Project Manager* and *Supervisor* with the right skills. Consider the various accreditation courses now available (see https://www.neccontract.com).

- Take care not to give or act on verbal instructions; follow the contract and put communications in a form that can be read, copied and recorded.
- Take care to use words such as 'acceptance' in properly managing the contract where the contract requires this.
- Appreciate that early warnings are a good and positive thing that can help in best practice contract management. They are entirely different from compensation events and cover very different things.
- Note that only the *Contractor* and the *Client* can change a term of the contract that exists, and the *Project Manager* cannot waive, amend or delete a project of another's contract.

NEC4: Practical Solutions, Second edition

Robert Alan Gerrard and Patrick Waterhouse
ISBN 978-1-83708-959-8
https://doi.org/10.1108/978-1-83708-956-720252002

Chapter 2
The *Contractor's* main responsibilities

26. Refusing to do what the Scope requires (clause 20.1)

QUESTION

The Scope requires our *Contractor* to prepare a breakdown of cost in a particular format, but it is refusing to do so. Does it have to do this? What provisions exist within the contract to enforce this?

ANSWER

Clause 20.1 is a very short, yet key, clause – the *Contractor* Provides the Works in accordance with the Scope. 'Scope' is defined in clause 11.2(16) as information that either specifies and describes the *works* or states any constraints on how the *Contractor* Provides the Works. This is either in the documents that the Contract Data states it is in or any subsequent instruction. Clause 14.3 allows the *Project Manager* to instruct a change to the Scope. The *Contractor* is therefore obliged to comply with the Scope and any subsequent instruction (see also clause 27.3). Ultimately, continued default of the *Contractor's* obligations could result in termination (clause 91.2 (R11)). The suggested approach would be to meet (perhaps by means of an early warning), explain the provisions within the contract and encourage compliance in order to avoid the ultimate sanction within the contract.

27. Who is responsible for design? (clause 21.1)

QUESTION

We are delivering a project under Engineering and Construction Contract (ECC) Option A, and the project is largely designed by the *Client*. The Scope stated that we, as the *Contractor*, should design the piling but was silent on the rest. The steelwork has been designed and calculated by the *Client's* structural engineer, but we are debating who should be responsible for the design of the fixing details.

ANSWER

The answer is simple. Under clause 21.1, the *Contractor* is only required to design the parts of the *works* that the Scope states it is to design. In this case, that is solely the piling. If the *Project Manager* now wishes you to design the fixing details, then a change to the Scope should be instructed, and this will then be a compensation event under clause 60.1(1).

28. Requesting more design information (clause 21.1)

QUESTION

We are using Option C of the ECC and are required to design and build a new sports complex for our *Client*. The Scope clearly states that the *Contractor* is responsible for all the design – which is not in contention. However, the *Supervisor* is insisting on seeing more and more of the drawings and design. This is incurring additional cost for us as it is insisting on paper copies of each drawing. Where do we stand with this?

ANSWER

One important initial comment is that the *Supervisor* has no power in the contract to request particulars of design. Clause 21.2 is the important clause in this situation. You are only required to submit particulars of the design as the Scope requires – if it says nothing, then that is exactly what you provide. If the *Project Manager* (not *Supervisor*) wishes to change the Scope under clause 14.3, and then requests particulars of the design, then you will be required to comply with this (clause 27.3). However, this change to the Scope will be a compensation event (clause 60.1(1)). The last sentence in clause 21.2 is very important and must be complied with – it states that you do not proceed with the work until the *Project Manager* has accepted. No timescale is mentioned in this clause, so we need to revert to the *period for reply* (clause 13.3). This needs to be built into your programme and assessed as part of the compensation event.

29. Proceeding with the work prior to acceptance (clause 21.2)

QUESTION

It is the first time that we have used the ECC, and we have opted for Option A. We followed the published NEC User Guides on the Scope, and specified the parts of the *works* that the *Contractor* is required to design. The project is fast track, and has £50 000 per day for delay damages. As a result of this the *Contractor* is progressing with the design and works very quickly. Our concern is that the *Contractor* is undertaking works prior to our acceptance.

Our question relates to a part of the design that has been submitted for acceptance. This has triggered thoughts in the *Client* team and, in hindsight, the original Scope needs changing. What do we do in this situation? Is this a compensation event?

ANSWER

What the contract says is that, assuming you requested the *Contractor* to submit particulars of its design (under clause 21.2), then it is not to proceed with the work in question until the *Project Manager* has accepted the design for that work. Failure to comply would be a default; continued default may result in termination of its obligation to Provide the Works (clause 91.2 (R11)).

This period of consideration is deliberately built into the contract to allow the *Project Manager* to review the design and, as in your case, potentially change your mind. The response period will be the *period for reply* (clause 13.3). The two reasons for not accepting (note that we do not use the term 'reject') are that it does not comply with the Scope or the applicable law. So, if you do not accept because you have changed your mind, then this would in turn be a compensation event (clause 60.1(9)). It is important to meet with the *Contractor* and explain these provisions within the contract.

30. Requesting the design of Equipment (clause 23.1)

QUESTION

I am the *Project Manager* on a large petrochemical project, and my internal health and safety manager is keen to see the design of earthwork support for deep trench excavations and a temporary footbridge required for the project. I cannot find a clause in the contract that allows me to request this.

ANSWER

The clause in this case is clause 23.1. Equipment under the ECC includes items provided by the *Contractor* and used by it to Provide the Works and that the Scope does not require it to include in the *works* (clause 11.2(9)). You can instruct (under clause 23.1) that the *Contractor* submit particulars of the design of an item of Equipment for acceptance. You can only 'not accept' if you consider that this will not allow the *Contractor* to Provide the Works in accordance with the Scope, the *Contractor's* design, which the *Project Manager* has accepted, or the applicable law. Unlike the design acceptance process for permanent works, there is no hold point here; the *Contractor* can erect its temporary works without awaiting the *Project Manager's* acceptance of the design.

31. Changing *key people* (clause 24.1)

QUESTION

We interviewed a *Contractor* as part of the tender process for an ECC project. The Site Manager proposed (listed as a *key person*) was highly experienced in the healthcare sector (which we are in). We have now heard that this person has obtained a new job and is about to leave. What provisions are there in the contract for dealing with this?

ANSWER

The qualifications and experience of this *key person* should have been detailed in Contract Data part two. Clause 24.1 requires that the *Contractor* submits the name, relevant qualifications and experience of any proposed replacement person for acceptance by the *Project Manager* – a reason for not accepting is that their relevant qualifications and experience are not as good as those of the person who is to be replaced. This does not bind you to accepting a replacement of less experience or qualifications, but you may do if you wish.

32. Not providing services and other things as stated in the Scope (clause 25.2)

QUESTION

We have a large highways project under Option B of the ECC. We specifically requested that the *Contractor* provided us with an office unit on Site with broadband and electrical connections. It has failed to provide these, and, as a result, we are incurring additional costs. What provisions are there in the contract for dealing with this?

ANSWER

Clause 25.2 requires that the *Contractor* provides services and other things as stated in the Scope. The same clause also states that any cost incurred as a result of the *Contractor* not so providing is assessed by the *Project Manager* and included as a negative adjustment to the amount due under clause 50.3 ('less amounts to be paid by or retained from the *Contractor*').

33. Missing a Key Date (clause 25.3)

QUESTION

We are the *Client* with a Key Date on our ECC Option C healthcare project. This is for the construction of a concrete plinth and first-fix electrics ready to take a cancer scanner from a specialist supplier with whom we have a direct contract.

The *Contractor* recently notified an early warning as, due to its delay, it anticipates missing the Key Date. How and when do we value this?

ANSWER

This is assessed when the *Project Manager* decides that the *Contractor* has not met the Condition stated by the Key Date, not before (clause 25.3). At this point, the *Project Manager* would assess the additional cost to the *Client* in carrying out work or paying an additional amount to Others in carrying out work. For example, this may include re-hire of a crane and storing the scanner in a warehouse. This should be cost incurred 'on the same project', so cannot be remote – it has to be directly incurred on the same project. The assessment must also be made within 4 weeks of the date.

34. Quality control of the supply chain (clause 26.2)

QUESTION

We are a *Client* and are about to embark on the construction of a high-security new-build prison and have selected the ECC. The choice of main *Contractor* is critical, which we will inevitably have control over during the tender process. Our concern is the Subcontractors and whether they are of the right calibre – what provisions are there in the ECC in this respect?

ANSWER

The ECC provides a level of control in this respect. There is no naming or nomination process. Clause 26.1 confirms that – whether the work is subcontracted or not – the *Contractor* retains responsibility. The next clause (26.2) is key to addressing your concern. This requires that, prior to the appointment, the *Contractor* submits the name of each Subcontractor to the *Project Manager* for acceptance. The *Project Manager* can only 'not accept' if it believes that acceptance will not allow the *Contractor* to Provide the Works. Reasons would need to be perhaps around health and safety or major concerns about quality assurance systems or vetting of staff. It is irrelevant what was stated during the tender process in terms of Subcontractors – the ECC only becomes operable once signed, so the *Contractor* will need to submit at this point. Clause 26.2 therefore provides a level of transparency – ultimately, it is the *Contractor's* responsibility. Failure of the *Contractor* to comply with this obligation may result in the *Client* terminating its obligation to Provide the Works (clause 91.2 (R13)).

35. Working with Others (clause 27.1)

QUESTION

We are a *Client* and have a complex engineering project under the ECC and have placed direct contracts with a number of suppliers. The civil engineering works are being undertaken by a main *Contractor* under Option A. We have several direct contracts with suppliers and other contractors with whom we require the main *Contractor* to liaise and coordinate its design. How does the ECC support this?

ANSWER

The other suppliers and *Contractors* that you have direct contracts with are classed as 'Others' under the ECC (clause 11.2(12)).

The contract refers to Others in a number of instances – in direct answer to your question, clause 27.1 is the most relevant – this requires the *Contractor* to obtain approval of its design from Others as necessary. It is also worth reviewing all of the clauses that refer to Others – these include clauses 25.1, 27.1, 60.1(5), 80 and 91.3.

36. No acceptance before subcontracting (clause 26.2)

QUESTION

We are the *Project Manager* on an ECC contract, main Option A. The *Contractor* has failed to seek our acceptance before subcontracting the installation of the highway safety fencing. The Scope requires the installation of the safety fence to be undertaken by a certified contractor under the National Highways Sector Scheme. The Subcontractor is not certified. When challenged, the *Contractor* says that the installer is a labour only contractor and as such does not require acceptance by the *Project Manager*. The *Contractor* also says that the posts and rails for the safety fence are paid for by the *Contractor* and the supplier of the fencing is certified under the Scheme.

ANSWER

The *Supervisor* should notify a Defect under clause 43.2 as this is a part of *works* not in accordance with the Scope as only a part of the Scheme may have been complied with. It looks like this organisation may not comply with what is a Subcontractor under clause 11.2(19), so clause 26 may not be relevant. Although we do not know how this is referred to on the Activity Schedule, it may be that the fencing works will fall foul of clause 11.2(29), in that the correction of the Defect may well delay following work.

FURTHER THOUGHTS AND TOP TIPS

- The Scope is a key document, which the *Contractor* is required to comply with. Make sure this is well understood and implemented.
- Ensure that the design submission and acceptance procedures are fully appreciated at the beginning of the project.
- Make sure that Subcontractors are accepted at the commencement of the project (before their appointment is made).
- The Scope should clearly define the *Contractor's* interaction with Others – on complex projects this is vital to define accurately.
- The impact of missing a Key Date is significant – ensure Key Dates are well planned and the Condition to be met is clearly defined.

Robert Alan Gerrard and Patrick Waterhouse
ISBN 978-1-83708-959-8
https://doi.org/10.1108/978-1-83708-956-720252003

Chapter 3
Time

37. Can a Completion certificate also be a payment certificate? (clause 30)

QUESTION

We have used the Engineering and Construction Short Contract (ECSC) on a small job that was completed a few months ago. We issued a Completion certificate (which was also the penultimate payment certificate) releasing half of the retention being held. The *Contractor's* bond people will not accept it as a Completion certificate for that job, making it difficult for the *Contractor* to get another bond for another job. Can a Completion certificate also be a payment certificate?

ANSWER

There is no such thing as a payment certificate in the ECSC; instead, the *Contractor* makes the assessment, applies for payment, and the *Client* decides what is due and pays (clauses 50 and 51). The requirement for a certificate for Completion to be issued by the *Client* is set out in clause 30.3. It does not have to be in any particular form, but it does have to state clearly on what date the *works* were completed and should be issued within 1 week of that date. It cannot also be a payment certificate because, as said, there is no such thing as a payment certificate. Note that in the ECSC there is no separate certifier as such, although the *Client* can delegate any of its actions to anybody, if that delegation complies with clauses 14.5 and 13.1. Whether or not this is accepted by anybody for any reason is not something we are able to comment on.

38. Has Completion been achieved? (clause 30.2)

QUESTION

Our query concerns certifying Completion within an Engineering and Construction Contract (ECC) project (Project 1) that consists mainly of a large-diameter water main, which the *Client* currently has beneficial use of. In accordance with clause 11.2(2), all the work within the Scope has been completed. However, recently a fitting on the water main (which was manufactured, supplied and installed by the manufacturer and/or its subsidiaries) started to leak. This necessitated the new water main being temporarily taken out of service while the fitting and surrounding section was cut out and replaced with a different product. After the repair to this Defect, the water main went back into service and the *Client* gained beneficial use of the same again. The faulty fitting had been awaiting transportation to an independent testing laboratory for inspection, analysis and a report into the likely causes and mode of failure. However, another fitting (again supplied, installed etc. by the same company) started leaking on another project (Project 2 – the same *Client* but a different *Contractor*). This fitting has now also been removed and sent for testing and a further report. As you can imagine, the failure of the same fitting on two different contracts has led to some cause for concern within the *Client's* organisation as to the potential for further failures and the possibility of further Defects. Contractually, how do things stand within Project 1? Obviously, the findings of the independent laboratory and report are the immediate concern; however, the *Contractor* may claim that a Defect occurred and, in accordance with clause 44, it was corrected. Therefore, while there are no further leaks (Defects), is the *Contractor* now entitled to Completion being certified?

ANSWER

It appears that the *Contractor* has been entitled to have Completion certified some time ago. The *Project Manager* should have certified Completion within 1 week of the *works* complying with the definition of Completion, and its failure to do so is a breach of the contract (clause 30.2). These matters may be Defects, which can and should be, or rather should have been, dealt with by the processes set out in clauses 43–46. In addition, as soon as the *Client* started using the water main or within 2 weeks of Completion (whichever happened earlier), the *Project Manager* should have certified take over (clause 35). Again, failure to do so within 1 week is also a breach of contract.

39. What goes onto the first Accepted Programme? (clause 31)

QUESTION

My query is in relation to the first Accepted Programme. We have a tender programme that we believe is included as a contract document but not referenced within Contract Data part two. The contract award was delayed by 12 weeks, and therefore the tender programme within the contract did not reflect the later contract dates. At the contract award, Contract Data part two stated 'To be submitted by the *Contractor* in accordance with the contract' (i.e. within 3 weeks). Following the contract award, we submitted a first programme for acceptance. This programme was the tender programme adjusted by the delay to the contract award but in line with the contract dates and original logic. This programme has not been accepted for the following reasons.

- The *Contractor's* plans which it shows are not practicable.
- It does not show the information that this contract requires.

The main issue relates to the *starting date*, which is currently in delay due to *Client* planning-approvals problems. The *Project Manager* is looking for us to submit a first programme, which reflects changes that have occurred since the Contract Date. The *Project Manager* appears to be looking for us to change our logic from the tender programme and show activities happening later because of the *Client's* delays.

Our query is whether this is reasonable, or should we be able to submit a programme in line with our original logic irrespective of changes that have happened since the Contract Date? As there are also several *Project Manager's* instructions already, which are going to be compensation events, any other situation would mean that the programme does not reflect the contract and the baseline at the outset. It should be noted that the programme that was submitted for acceptance was also lacking resources and other matters, which we realise need to be added for acceptance. We would be grateful for any guidance on this matter prior to us arranging an early warning meeting with the *Project Manager*.

ANSWER

Your tender programme is only the first Accepted Programme if it is referred to in Contract Data part two. Otherwise, it is not the first Accepted Programme, and we cannot comment on the status of it with such limited information; the incorporation of a programme into the contract could have quite dramatic implications.

Putting the above to one side, you need to produce a compliant programme that reflects things as of the Contract Date; this will line up with the other dates and your obligations. Things that happen after that, such as *Project Manager* instructions to change the Scope, other *Client* delays and so on, will be dealt with in the order they are known about and not on the first Accepted Programme. You could not predict these sorts of problems, nor of course does the contract expect you to.

40. Incorporating a programme into Subcontract Scope (clause 31)

QUESTION

We are a main contractor using Engineering and Construction Subcontract (ECS) main Option B. The Subcontractor is providing temporary traffic management, and the main item is to provide 4 km single-lane overnight closures on a dual carriageway highway. The quantity of closures that are needed has increased. The Subcontractor has notified a compensation event, requesting an enhanced rate/change to the original Bill of Quantities (BoQ) rate. The Subcontractor is using clause 60.1(1) as the basis of its entitlement, saying that any deviation from the prescribed programme and resources therefore constitutes a change to the Subcontract Scope. The reason for this is that the *Contractor* has identified within the Form of Agreement the appendix where the Subcontract Scope is contained, and within this appendix we have included a programme and histogram. Within Subcontract Data part two we have: 'If a programme is to be identified in the Subcontract Data – the programme identified in the Subcontract Data is identified in Subcontract Scope Appendix A item 4.'

Can the Subcontractor rely on the inclusion of the two documents – the programme and histogram – within the appendix as the Subcontract Scope? And as such is it entitled to a compensation event under clause 60.1(1)? We consider that a change to the Subcontract Scope at the risk of the Subcontractor or as a consequence of its proposals is not a compensation event. During the currency of the *subcontract works* the Subcontractor failed to submit a revised programme as required under clause 32. Does this further impact on the Subcontractor's entitlement to a compensation event as it has failed to identify the effects of the compensation event?

Finally, the Subcontractor has only recently identified its entitlement to the compensation event pursuant to clause 60.1(1), which is several months after Completion. It had previously claimed an entitlement under clause 60.4 but was unable to demonstrate that the Defined Cost of its resources had changed. Does this change in argument/direction have the same effect of clause 61.7 and effectively time bar any compensation event claim? Note that the subcontract had sectional Completion Dates, and as such certain sections are now beyond the *defects date*.

ANSWER

If you have somehow included a programme within the Subcontract Scope, then if you change that programme that becomes a compensation event because you are changing the Subcontract Scope. And that compensation event will be assessed not based on the rates in the BoQ but on the actual increase in Defined Cost. So, it is not a case of enhancing a rate – the Subcontractor must demonstrate the Defined Cost. That may be more or may be less than it has in the BoQ, but it still must be demonstrated. It is the Subcontractor that is supposed to provide the programme, not you. It is for you to accept or not accept it. It is its programme for its *subcontract works*. We suggest you look at clauses 11.2(1) and 31 and 32 for more information on this, and it is in the Subcontract Data rather than the Subcontract Scope because programmes will inevitably change! So, your mistake was saying at the entry in the Subcontract Data to look in the Subcontract Scope for the programme. Because now, whenever it changes, it is a compensation event. The time bar in clause 61.3 does not apply to clause 60.1(1) because it is a compensation event that you were supposed to notify, not the Subcontractor (clause 61.1). We cannot see how the time bar in clause 61.7 can apply when you have not yet got near the *defects date*. The answer to that will be in what you have set the *defects date* up to be in the Subcontract Data. So, yes, it is a compensation event, but the Subcontractor still has to demonstrate what effect it had on Defined Cost, just as it does with any other compensation event.

41. Programme float and time risk allowances (clause 31)

QUESTION

On an ECC Option A contract, in presenting the clause 31 programmes, the *Contractor* has not defined any float or time risk allowance; rather, it has included a terminal risk allowance. With reference to the critical path, we consider that not only does this preclude the ability to monitor risk/residual risk as the *works* progress, but it also prevents assessment in terms of the sufficiency of the operational resource and time allowances and the deliverability of the programme generally. It is worth noting that the critical path constitutes complex bridge construction works over an environmentally sensitive water course. Could you advise on the resolution of the issue, regarding the *Contractor's* position that there is no float within the critical path, and an assessment that a programme with no float on the critical path is impractical/unrealistic and, therefore, cannot be accepted?

An early warning meeting cannot fully resolve the matter (the *Contractor* has already stated that, due to its commercial approach on the job, there is no float/risk provision in the critical path operations), and there only appears to remain a completely unsatisfactory situation whereby these mutually exclusive positions lead to an impasse, undermining the programme as a management tool.

ANSWER

The *Contractor* is required to show float and time risk allowances on its programme (clause 31.2). If the *Contractor* does not do so, then the *Project Manager* must notify the reasons for not accepting under clause 31.3 (second bullet). A programme without any time risk allowances or float, or where all the time risk allowance is at the end, is also simply not practicable to achieve and should therefore not be accepted under the first bullet of clause 31.3.

However, regarding the critical path, the *Contractor* is correct that there simply cannot be any float within that – by its very definition it is the path that is critical. In fact, one of the accepted definitions of the critical path is that it has zero float, or, to put it another way, it is the quickest period that the work can be carried out in. Generally, now, planning software will identify the critical path and float for the remaining items once the programmer has entered the time periods for each activity (including any time risk allowances for that activity) and the links between the activities (either logic or resource driven). However, the *Contractor* is best advised not to lump all its time risk allowances together at the end of the programme. These should be included for each activity that needs them (not all will). The amount of these will depend on many factors, including the nature of the activity and the time of year it is being carried out. That way, the programme shown will be achievable (i.e. practicable). The operation of the compensation event assessment provisions ensures that time risk allowances remain the *Contractor's*, and the *Client* cannot use them when assessing a compensation event.

42. What becomes of a tender programme? (clause 31)

QUESTION

As part of a quality assessment, we asked tenderers for a resourced programme submitted from the stages of design to Completion. As you might expect, these varied in quality. Contract Data part one stated: 'The *Contractor* is to submit a first programme for acceptance within 4 weeks of the Contract Date.' We have a few questions on this scenario.

1. In awarding the contract, are we bound to accept the tender programme as the clause 31 programme?
2. Are we entitled to request a new clause 31 programme based on the actual Contract Date?
3. At award of contract, the Contract Date was 2 weeks later than had been assumed by the tender programme. Is the *Contractor* entitled to have Key Dates and/or the Completion Date moved forward by 2 weeks?

ANSWER

The answer to Question 1 is going to depend on the status of the programme issued at the tender stage. If it was referred to by the *Contractor* in Contract Data part two, then it becomes the first Accepted Programme (clause 11.2(1)); the reference in Contract Data part one is irrelevant because the 'if' statement does not then apply.

Regarding Question 2, we are not sure that in practice this is particularly relevant, because the programme must be revised in any event at regular intervals (clause 32).

Regarding Question 3, the Contract Date has no relevance to a programme issued for acceptance – it is the *starting date* and *access dates* that are important (clause 31.2). These are the dates that were in your Contract Data unless they were perhaps amended in tender correspondence.

43. What do we do if the programme shows a wrong date? (clause 31)

QUESTION

On an ECC contract, the *Contractor* is asking for acceptance of a programme that shows the effects of a *Project Manager's* instruction, which is a compensation event affecting both the Prices and planned Completion. I believe the *Contractor* has not assessed this correctly, and the assessment is not currently agreed between us. I have not accepted the submitted programme because the programme shows the *Contractor's* critical path delay as 53 days, whereas my assessment is 29 days. The programme thus indicates an incorrectly revised Completion Date. Is this a reason for not accepting under clause 31.3?

I am looking at the second bullet of clause 31.3 here and considering the direction of the second bullet of clause 32.1 and the first bullet of clause 32.1. I do not understand how I can accept a programme that shows something that is not agreed. This compensation event is significant in value, and I believe it will take some time to settle (maybe by means of the *tribunal*), as we have reached an impasse.

After this event, we have had a further significant compensation event involving a delay, which is agreed in both cost and time. I am therefore required to accept a further programme to reflect this change. How should I progress this if the previous matter is not agreed?

ANSWER

There are two separate dates that must be shown on the programme for acceptance. The date for planned Completion is one such date (second bullet of clause 31.2). This is the date on which the *Contractor* plans to achieve Completion. It will be at the end of the critical path and all the activities will, directly or indirectly, link to it. It can be before, on or after the Completion Date. If it is the latter, the *Contractor* is admitting it is going to finish 'late'.

The Completion Date is the other such date (first bullet of clause 31.2). This is the date by which Completion should be achieved. This is not linked to anything else on the programme and can only be moved as a result of a compensation event (clause 63.5), accepting Defects (clause 45.1) or agreed acceleration (clause 36.3).

You need to make sure that you understand the difference between these two and do not confuse them. You also need to understand that terminal float (i.e. the period between planned Completion and the Completion Date) remains the *Contractor's* responsibility (plus or minus) and cannot be used to alleviate the effect of a compensation event (clause 63.5).

Regarding the compensation event, the time for 'settlement' is irrelevant to you. If you have concluded that the *Contractor's* quotation is incorrect, clauses 64.1 and 62.3 require that you make your own assessment of both the money and time consequences of the compensation event. Clause 64.3 requires that you do that within 3 weeks. Once you notify the *Contractor* of this assessment, the compensation event is implemented under clause 66.1. You then include in your payment certificate the amount you have assessed (Options A or B) when these works are complete or add it to the target (the Prices, in Options C or D). You also extend the Completion Date by the amount you consider it is due. That is the end of your involvement.

The *Contractor* can (possibly negotiate with the *Client* through the *Senior Representatives* or) refer the matter to the *Adjudicator* to get it changed. If the *Contractor* believes it will take 53 days to deal with this event, it is entitled to put that on its programme for acceptance, and therefore move the date for planned Completion by that amount. That will not move the Completion Date, and if you accept that programme it will not mean you are accepting that the Completion Date should move by that, or any other, amount. On the other hand, if you do not think it will take 53 days to deal with the event, then you must not accept the *Contractor's* programme because it does not represent the *Contractor's* plans realistically (clause 31.3, third bullet). Of course, if you are wrong on that, then eventually the *Contractor* will be likely to win the argument if it goes to adjudication.

44. Programme not being responded to by the *Project Manager* (clause 31.3)

QUESTION

As a *Contractor* we consistently find that our programmes are not responded to by the *Project Manager* on our ECC projects. Can you offer advice as to how we should best deal with this and how we should protect ourselves?

ANSWER

Unfortunately, this is not uncommon! The good news is that clause 31.3 in the ECC does provide expressly for this.

It is worth noting that the *Project Manager* should 'act as stated' (clause 10.1). This means that it should either 'accept' or 'not accept' within 2 weeks of programme submission. The reasons for not accepting are listed in clause 31.3.

In the ECC, if the *Project Manager* does not respond within the time allowed, the *Contractor* may (optional) notify (clause 13.7 in a separate communication) of its failure. If the failure continues for a further week, it is treated as acceptance by the *Project Manager* – effectively 'deemed accepted'.

The very existence of this provision should hopefully ensure that the *Project Manager* does indeed respond. Teams that work effectively under the ECC often agree to meet before the submission deadline, review the programme and agree informally to the revised programme. This is then followed with a clause 32.2 submission and clause 31.3 acceptance. It is also worth noting that clause 10.2 requires that the *Project Manager* acts in a spirit of mutual trust and co-operation, so it should be proactively looking to work with the *Contractor* and agree revised programmes.

45. Problems with a programme submitted at the tender stage (clause 32)

QUESTION

We are at present engaged on a major construction project using ECC Option A and seek clarification on a point of procedure regarding the programme. All tenderers submitted a programme as part of their tender, as they were required to do. Following evaluation of the tenders, the *Client's* agent (this role includes acting as *Project Manager*) recommended award to contractor X. No comment was made at the time by the *Client*/agent regarding contractor X's programme, and there was no notification of acceptance or reasons for non-acceptance of this programme. The contract was formally awarded by the *Client* to contractor X (the *Contractor*) in early October.

The *Project Manager* advises that the *Contractor* submitted its first programme for acceptance in January. The *Project Manager* wrote to the *Contractor* 13 days later with comments that required attention prior to the *Project Manager's* acceptance. The *Contractor* responded, claiming that the *Project Manager's* response had neither accepted nor notified reasons for not accepting, stating that one of these should have been due within 2 weeks of submission and it therefore deemed the programme submitted in January to be the Accepted Programme. The *Contractor* then submitted an updated programme in February taking cognisance of the comments and requirements sent by the *Project Manager* earlier. The *Project Manager* wrote to the *Contractor* more than 2 weeks later, indicating that the programme submitted in February was not accepted for all four clause 31.3 reasons. The *Contractor* maintains that there is an Accepted Programme – probably referring to its 'deemed' Accepted Programme.

We understand that if a programme was required to be submitted with the tender, the programme is identified in the Contract Data at the Contract Date. In this situation, an Accepted Programme already exists, and no amount can be retained under clause 50.5. We also understand that once the programme has been accepted, it becomes the Accepted Programme. A commentary on the ECC says that

> the contract does not deal expressly with the position if the *Project Manager* finds the programme identified in the contract data unacceptable. It is the Accepted Programme. Presumably this is a matter to be dealt with before the award of the contract.

None of the above considers the possibility in which the *Project Manager* neither accepts nor notifies reasons for not accepting, as appears to be the case in our project. We would welcome any comments and advice you may be able to offer, particularly regarding whether an Accepted Programme (or deemed Accepted Programme) exists, and to which one would this relate. Also, if this relates to the one submitted with the tender, how should any deficiencies in information that would have led to it not being accepted had it been properly assessed at the time be dealt with now?

ANSWER

The first thing you and your *Contractor* need to understand is that the programme is not a single document but a continually evolving one. The contract requires the *Contractor* to submit revised programmes at regular intervals – note the final bullet of clause 32.2 and the entry you have in your Contract Data part one. We always recommend that the interval should be no longer than monthly, ideally shorter. Once each programme has been accepted, it becomes the Accepted Programme until the next one is issued and accepted (clause 11.2(1)). And so, the cycle goes on.

The status of the programme submitted with the *Contractor's* tender depends on how it has been referred to within the Contract Data, so unfortunately, we disagree with the quoted suggestion above that its status is unclear – nothing could be clearer. If the programme has been referenced in Contract Data part two by the *Contractor*, then it automatically becomes the first Accepted Programme – the wording of clause 11.2(1) is very clear on that. Of course, it then gets superseded by future programmes once they are accepted. If it is not referred to in the Contract Data, it is not an Accepted Programme. If the tender assessment team is not happy with this programme, then do not accept the tender, or put up with it or negotiate (if procurement rules permit).

Looking at what has happened, it seems to us that there has been a catalogue of errors on both sides. Firstly, it seems that the *Contractor* was very late in submitting its first post-contract programme for acceptance. And the *Project Manager* has also not done what should have been done. If the tender programme was not the Accepted Programme, for the reasons in the previous paragraph, the *Project Manager* should have been retaining 25% of the Price for Work Done to Date in assessments until a 'correct' programme (clause 50.5) was submitted. Even if the tender programme was the Accepted Programme, the *Project Manager* should have been chasing for another one and instructing one if need be (first bullet of clause 32.2). Once the *Project Manager* receives a programme, it has 2 weeks to say 'yes' or 'no', and if it says 'no' it has to give reasons (clause 31.3). If the *Project Manager* has some comments it wants to discuss with the *Contractor*, it is a bit too late to write after 13 days. Why not just talk to the *Contractor* about the programme and discuss any concerns? And then the *Project Manager* failed to reply within the 2 weeks, which is a compensation event (clause 60.1(6)). But then the *Contractor* got it wrong, as well. Clause 31.3 deals with the situation whereby the *Project Manager's* lack of response after receiving a reminder might lead to a programme that is treated as accepted. As previously mentioned, the *Contractor* also has the protection of notifying a compensation event. But it is now too late to do that (clause 61.3). The *Contractor* then submitted a new programme for acceptance – absolutely correct. But the *Project Manager* has not accepted that for all four reasons in the contract. That means it is either an appalling programme or the *Project Manager* is perhaps covering against all eventualities, or somewhere in between. None of those options bodes well for the future.

The present position is that you either have no Accepted Programme at all or the (well out of date) one in the Contract Data is the Accepted Programme. That can be in neither Party's interest. So, our first piece of advice is for the *Project Manager* to sit down and talk constructively with the *Contractor*. Neither has covered itself in glory so far, so both should approach those discussions with a degree of humility, and, above all, with a positive desire to get a programme that both sides can live with. That is far better than having no Accepted Programme at all or, worse still, an out-of-date, and therefore useless, one. And, in any event, acceptance does not impose any liability on the *Client* or the *Project Manager* (clause 14.1). A further point is that until the *Contractor* starts issuing programmes for acceptance within the periods in the contract, and the *Project Manager* accepts them, the *Project Manager* is required to assess all compensation events, using what it believes the Accepted Programme should be (clauses 64.1 and 64.2), and it must do that within 3 weeks (clauses 64.3 and 62.3).

46. Lack of acceptance of programmes (clause 32)

QUESTION

We are a Subcontractor on a contract using the ECS. We are struggling to get acceptance of any of our programmes and have to date submitted ten programmes for acceptance by the *Contractor,* none of which has been accepted. Each time our programme is not accepted, the *Contractor* changes the *subcontract access dates*, and we resubmit the programme with the new access dates for some sections of the work, and again it is not accepted. The work is progressing ahead of our original programme. Can this situation carry on until the end of the project, where the *Contractor* keeps on not accepting our programme, or is there a mechanism in the ECS where it must accept a programme?

ANSWER

You are required to submit programmes for acceptance as set out in clause 32.2. Once you have done so the *Contractor* has 3 weeks to accept or not accept it (clause 31.3). If the *Contractor* does not do so within that period, then that is a compensation event (clause 60.1(6)), and you will be compensated for the effect that the event has had on the Defined Cost and the programme. If the *Contractor* does not accept it, the *Contractor* must give reasons (clauses 31.3 and 13.4). You are required to resubmit the programme considering those reasons (clause 13.4). If the reason is listed in the contract (clause 31.3), then that is the end of the matter (clause 13.8). If the reason is not one of those listed in the contract, then that is also a compensation event (clause 60.1(9)). The problem you may have with these compensation events is that you have only got 7 weeks from when you knew about them to notify them as a compensation event, otherwise you lose all your rights to additional time and money (clause 61.3). In addition, if the *Contractor* issues an instruction changing when access will be provided, each of those instructions will be a compensation event (clause 60.1(2)).

47. Showing non-implemented compensation events on a programme (clause 32)

QUESTION

Under an ECC Option B contract, we have several instances in which the *Project Manager* has instructed a change to the Scope, but the compensation event has yet to be implemented (as the quotation is either outstanding or not yet agreed); the effects on the programme have therefore not been accepted. The *Contractor* has submitted its latest clause 32 programme for acceptance and included the programme effects for these compensation events, which are not yet implemented.

As the *Client*, we view this as incorrect as we have not yet notified acceptance of these compensation events. Is this position correct?

We are concerned that if we do accept a programme that includes the effects of implemented compensation events, we may by default be agreeing to the time effect of the compensation event when this is not the case. However, we appreciate that the *Contractor* is keen to reflect the actual position within its programme regarding instructed work that has been completed before the compensation event has been implemented – thus to ensure it shows this work as as-built on the programme.

ANSWER

It is important to understand that a programme submitted for acceptance must show both the Completion Date (contractual) and planned Completion (likely) (clause 31.2). The latter is linked through the activities in the programme, the former is not. The latter can move for all sorts of reasons, the former can only move because of a compensation event. If the *Contractor* knows something is going to happen, it must show its effects on the programme straight away, whether it is a compensation event or not, and that may or may not affect planned Completion. That is because it must show 'the order and timing of the operations which the *Contractor* plans to do' (third bullet of clause 31.2).

The practical effects of events, therefore, including compensation events that are known about but not yet implemented, have to be shown on the programme. So, planned Completion may move, but the Completion Date does not and cannot at that stage. Once the compensation event is implemented, then the Completion Date is moved, and that movement is shown on the programme. If the *Contractor's* programme follows those rules, then the *Project Manager* should accept it, if it is compliant in other respects. Acceptance will not infer or imply anything about the compensation event. And it will not take any responsibility from the *Contractor* (clause 14.1).

48. Take over and its effect on liability (clause 35)

QUESTION

What effect does certifying take over for part of the *works* have on the *Contractor's* insurances/liability for that part? We have an ECC Option A contract for environmental improvement works (new footways, lighting etc.) in which vandalism and anti-social behaviour have increased in recent months. As a result, the *Contractor* is keen to protect itself at the earliest opportunity.

ANSWER

Take over of all or any part of the *works* occurs when the *Client* starts using that part (or all), unless that use is for a reason stated in the Scope or to suit the *Contractor's* method of working (clause 35.2). One of the *Client's* liabilities listed in clause 80.1 is loss of or damage to any part of the *works* that have been taken over (sixth main bullet). Up until take over, the *Contractor* is responsible for this (clause 81.1) and must insure it (clause 83.2). Take over is therefore important because once any part of the *works* is taken over, the *Contractor* is not liable for damage unless caused by it. Clause 83.3 requires the *Contractor* to maintain insurances until the issue of the Defects Certificate, which will be after take over.

49. A *Client's* liability? (clause 35)

QUESTION

I am the *Project Manager* on a project using ECC Option C. Completion was achieved on 4 June, and the *Client* took over the works on 11 June. However, a few days later, due to heavy rain, topsoil was washed off an area of the embankment slope, which also took with it landscape planting. The *Contractor* states the scheme is handed over. We state it is a Defect as the topsoil layer should not have failed.

ANSWER

The first place we would look in the contract is the set of *Client's* liabilities in clause 80.1. The sixth main bullet says

> … loss of or damage to the parts of the *works* taken over by the *Client*, except loss or damage occurring before the issue of the Defects Certificate which is due to … a Defect which existed at take over … or … the activities of the *Contractor* on the Site after take over.

If it was the case that there was no Defect that existed at take over, or the *Contractor* has carried out no untoward activities after take over to cause this, then this matter would be a *Client's* risk. The *Contractor* would have done that which it was obliged to do, which is to Provide the Works in accordance with the Scope.

A Defect is a part of the *works* that is not in accordance with the Scope (clause 11.2(6)). Were tests or inspections carried out to verify whether the *Contractor* had done that which the Scope stated it should do? Was the Scope sufficient in itself – that is, did the *Contractor* do right but the brief was wrong in the first place? How can you now be sure that there was indeed a Defect in how the *Contractor* had provided the *works*? How can you be sure that the constructed works should not have failed?

We would imagine it is going to be extremely difficult to prove a Defect existed at the time of take over without any evidence of tests or inspections at the time. If somehow it can be proved, then this would be a *Contractor's* risk under clause 81.1. Are there other areas of similar topsoil and planting that demonstrate the Scope has not been followed? Maybe the Scope was correct, and the *Contractor's* work was acceptable, but it was unfortunate timing that this part of the *works* had not had a chance to settle before the damaging rain – this is not a *Contractor's* risk after take over.

50. Damage by Others after take over (clause 35.2)

QUESTION

We are using ECC Option C and would like to know what is the correct (contractual) position in respect of damage caused by Others? For example, in the instance that we, as the *Contractor*, had completed our *works* and subsequently handed over an area of the completed *works* to another contractor (which is not part of our contract) by request of the *Project Manager*. What is the position of the *Contractor* when the *Project Manager* states that the *Contractor* was the Principal Contractor and therefore holds responsibility for damage caused by Others (irrespective of the handover documentation that was issued to the other contractor from the Principal Contractor clarifying the state of the completed *works*). You should note that the contract makes no provision in allowance for interim handovers to Others and that there is no provision or allowance for resources in the Prices for monitoring those works carried out by Others.

ANSWER

We are afraid that there is no such thing as 'handover' in the ECC, so it is difficult to comment on this, other than to reflect on the correct wording/processes in the ECC. NEC contracts are not the same as other contracts, and everybody needs to take care to use the correct principles and terms when managing them or confusion might arise.

If the *Client* wants to use a part (or all) of the *works*, either for its work or for the work of Others (as here), then it takes over all, or part of, the *works* (clause 35.2) (we assume from what you say that neither of the two exceptions in the clause's bullets applies in this case). The *Project Manager* is then required to certify the date for take over within 1 week (clause 35.3). That is what you should have got the *Project Manager* to do, rather than using the term 'handover'.

The sixth main bullet of clause 80.1 makes it clear that the *Client* is liable for damage to the *works* once they are taken over, not the *Contractor* (clause 81.1). And you are no longer required to insure the *works*, because the first sentence of clause 84.2 makes it clear that the *Contractor* only has to insure against its risks. The fact that you are the Principal Contractor would likely be irrelevant, but we cannot give legal advice.

51. Where use of the *works* may not constitute take over (clause 35.2)

QUESTION

On an ECC Option A contract, the *Client* wishes to use a part of the *works* for a number of weeks but considers that the *works* are not complete, and believes and is concerned that such use will automatically trigger Completion for that part of the *works*, the beginning of the countdown to the *defects date* for that part of the *works* as well as having insurance consequences for the *Client*. To that end, the *Client* has requested the *Contractor* to confirm in writing that such use does not constitute take over. Such confirmation is required by the *Client* as a precondition of the *Client* using the *works*. If the *Contractor* does not provide such confirmation, the *Client* will be obliged to use an area that is outside the Site.

At the request of the *Client*, a 'Statement of Readiness of Use' has also been provided by the *Contractor* as a second precondition of the *Client* using the *works*. The statement provided confirms that the *works* are complete but for two items, which in no way prevent or hinder the immediate and full use of the *works* as requested by the *Client*.

The *Contractor's* view is that such use of the *works* by the *Client* does constitute take over of that part of the *works* under the contract, but that that does not automatically equate to Completion either of the whole of the *works* or, as the *Client* believes, of that part of the *works*. The *Contractor* does not dispute that it must complete the *works* whether take over has been certified. Note that clause 11.2(2) has been amended to the following.

> Completion means in any event a state in which the Works are complete in all respects and free from apparent Defects, save for any minor items of incomplete works or minor Defects the existence, Completion or rectification of which in the opinion of the *Project Manager* would not prevent or interfere with the use and enjoyment of the *works*; provided where it is expressly stated in any provisions of the Scope that the testing, commissioning, regulation or adjustment of any mechanical or electrical services is to be completed before practical Completion of the works shall not be considered to be practically completed until the same is done as the Contract Documents require.

ANSWER

The *Client* appears to have misunderstood the contract. Completion and take over are two entirely different concepts in the ECC. Completion is about what state the *works* are in – that is, do they meet the requirements set out in clause 11.2(2)? Completion is what the *Contractor* must achieve before the Completion Date. Take over is about who is using the *works*, as set out in clause 35. Once the *Client* starts using the *works*, the *Client* takes them over. The *Client* is entitled to start using all or any part of the *works* before Completion, and if it does so, it has taken over those parts of the *works* (clause 35.2). However, Completion will still not be achieved until the *works* have met the tests set out in clause 11.2(2). At take over, the *Client* becomes liable for any loss of or damage to the *works* (sixth main bullet of clause 80.1). As these are no longer the liability of the *Contractor*, the *Contractor* is no longer required to insure them (first sentence of clause 83.2). However, the *defects date* is related to the date of Completion, not the date of take over (see the relevant entry in Contract Data part one). If the *Client* does take over early (i.e. before Completion and before the contractual Completion Date), that will be a compensation event. However, if the *Client* takes over when the *Contractor* is already late (i.e. before Completion but after the Completion Date), that is not a compensation event. See the wording of clause 60.1(15). It is for the *Project Manager*, not the *Contractor*, to decide when Completion and take over have occurred and to certify them (clauses 30.2 and 35.3). The *Project Manager* does so by applying the meaning to those terms set out in the contract. There is no such thing as a certificate of readiness of use in the ECC; you are not required to provide it, and it means nothing.

As a final point, it appears to us that the *Client's* changes to clause 11.2(2) are a recipe for an argument and are too subjective. The drafting seems to introduce new defined terms and does not properly use the identified terms. What exactly is a 'minor' item of work? And what does the reference to 'practical Completion' mean? The contract was designed to remove subjective judgements such as these, and replace them with objective decision-making, by defining what is meant by Completion. This change has reintroduced the uncertainty that vague words such as 'substantial Completion' and 'practical Completion' bring.

52. The interaction between take over, Completion and delay damages (clause 35.2)

QUESTION

We are currently engaged on an ECC Option A project including Options X5 and X7. Option X5 defines 27 individual Completion Dates and Option X7 allocates delay damages to 26 of the 27 Completion Dates.

The Scope on this project is silent on what constitutes Completion for any of the individual Completion Dates, and a dispute has arisen over when Completion was achieved. As the *Contractor*, we believe that Completion was achieved at the latest when the *Client* provided access to Others (contractors) to start their work. Are we correct with this view in the absence of any other Completion criteria in the Scope?

Several of the individual Completion Dates are based on providing a complete set of eight foundation bases. The *Client* has started erection works prior to all eight bases being complete but has refused to provide Completion for the bases that it has effectively taken over early. How should the *Project Manager* have dealt with this situation?

What are the obligations of the *Project Manager* regarding deduction of delay damages? Is the *Project Manager* obliged to deduct delay damages in full unless the Completion Date(s) are altered by an agreed compensation event? What happens with delay damages in a scenario such as that of the foundation bases – should the *Project Manager* levy delay damages in full?

If the *Project Manager* fails to deduct delay damages from an interim assessment, is the *Project Manager*/*Client* time barred at any stage from subsequently revisiting the applicability of delay damages? We have received notification from the *Project Manager* requesting payment of delay damages because they were not deducted in every monthly assessment when they should have been.

The wording used in Option X5 for each of the Completion items is slightly different from the same headings used for the delay damages under Option X7: for example, under Option X5 the Completion Date is defined as a date for Array A but under Option X7 it is defined as a sum for Array 1. Is it imperative that the wording in Options X5 and X7 match for the provisions on interim Completion and delay damages to apply?

ANSWER

Completion of a *section* of the *works* is dealt with in the same way as Completion (clause X5.1). In the absence of anything in the Scope, the definition of Completion is set out in the final part of clause 11.2(2). It is when 'the *Contractor* has done all the works necessary for the *Client* to use the *works* and for Others to do their work'. It appears the *Contractor* has achieved that by the fact that Others (i.e. these other *Contractors*) are carrying out their work.

However, you must also remember as well as Completion there is also take over to consider. Completion is about the state that the *works* are in (clause 11.2(2)), whereas take over is about who is using the *works*. The *Client* takes over the *works* (or any part of the *works*) when it starts using them (clause 35.2), and that would include getting Others to use them. The *Project Manager* must certify take over as well, within 1 week of the use starting (clause 35.3). So, on the information you have given us, take over must have occurred, regardless of whether Completion has been achieved, and the *Project Manager* is in breach by not certifying as such.

You will see in Option X7 that delay damages cease on the earlier of Completion or take over happening. Therefore, delay damages cannot be applied to these *sections*.

In addition, the *Client* can take over part of the *works* or part of a *section* of the *works* at any time. For example, in your case, the *Client* could start using six out of the eight bases in a *section* (clause 35.2), in which case those six bases are then taken over. In that case, the delay damages for that *section* are reduced in the ratio to the amount of benefit that the *Client* has got from the six as opposed to the eight bases (clause X7.3).

The *Project Manager* (not the *Client*) deals with the deduction of delay damages as part of its assessment (third bullet of clause 50.3). The *Project Manager* is obliged to deduct them when they become payable. If the *Project Manager* does not do so, that is an error, but this can be corrected in any later assessment (clause 50.6). There is no time bar in the contract for this to be done.

Obviously, the *sections* should have the same name in Options X5 and X7. If they do not, that can cause problems, but that is a legal matter on which we are unable to offer advice. In the scenario you have set out, it is unlikely that any delay damages would be payable, if the *Client* started using the bases by the Completion Date for those bases. The fact that the *Project Manager* has not done what it should (i.e. certifying Completion and/or take over) does not mean that you are liable.

53. No revised programmes submitted (clause 32)

QUESTION

What happens where a *Contractor* refuses to submit revised programmes in accordance with ECC clause 32.2?

ANSWER

This would be extremely frustrating in terms of contract management but also a breach of contract by the *Contractor*, but what losses would the *Client* be able to demonstrate it suffered? That said, this would not render the contract impossible to perform and it follows that the *Project Manager* would be obliged to assess all compensation events as the *Contractor* has not submitted a programme that the contract requires it to submit (clause 64.1). We cannot think how you can force an unwilling *Contractor* to submit a revised programme, you just apply the sanction.

54. Take over (clause 35.3)

QUESTION

Must the *Project Manager* issue a take over certificate once the *Client* takes over a part or the whole of the *works*?

ANSWER

Yes, clause 35.3 clarifies that the *Project Manager* certifies the date on which the *Client* takes over any part of the *works* within 1 week of that date.

55. Mobilisation and demobilisation plans (clauses 33 and 34)

QUESTION

What is the relevance of the mobilisation plan and demobilisation plan in the Facilities Management Contract (FMC)?

ANSWER

Clause 33 covers the requirements of the mobilisation plan and clause 34 the demobilisation plan. As it suggests in the titles, the first plan is concerned with setting things up properly and the second with closing the FM services down. There could be periods of study, collecting data, observing and handover from the existing service provider, dealing with staff transfers, procuring Equipment, obtaining licences, putting ICT systems in place, and checking that they work, engaging Subcontractors and suppliers, and so on. The demobilisation plan is basically the reverse of these.

56. Can the programme ever be automatically deemed accepted? (clause 31.3)

QUESTION

We note the provisions for deemed acceptance of the programme in clause 31.3 but if the *Contractor* doesn't notify the failure, can it ever be deemed accepted?

ANSWER

If the *Project Manager* does not reply to the submitted programme within the 2 weeks allowed in clause 31.3, then the obligation still sits with the *Project Manager* to respond as soon as it can. In the meantime, a compensation event occurs for the late reply under clause 60.1(6) and the 'treated as accepted' provisions can only occur if the *Contractor* notifies the *Project Manager* of the failure. Note that the contract does not use the word 'deem'.

FURTHER THOUGHTS AND TOP TIPS

- Make sure that the state of Completion is properly thought about and defined in the Scope when preparing the tender documents.
- Think of the programme requirements in clauses 31 and 32 as being a set of minimum requirements – there is nothing to stop inclusion of more than the minimum if the Parties agree that it would help the contract management.
- With the first programme in particular, the *Contractor* should get the person(s) who prepared this document to spend some time walking the *Project Manager* through it, to understand better the logic and any pinch points that concern the *Contractor*.
- Make sure that take over, testing and Completion are understood, properly reflected in the tender documents, planned out properly and clearly shown on the Accepted Programme.

Robert Alan Gerrard and Patrick Waterhouse
ISBN 978-1-83708-959-8
https://doi.org/10.1108/978-1-83708-956-720252004

Chapter 4
Testing and defects

57. Frustrated *Supervisor*! (clause 41.1)

QUESTION

I am a frustrated *Supervisor* under an Engineering and Construction Contract (ECC) Option A contract! I was not involved in drafting the original Scope but understand that no specific tests or inspections were stated. What tests and inspections should the *Contractor* undertake and how can I instruct additional tests and inspections?

ANSWER

The frustration is appreciated and, worryingly, not uncommon. Drafting the Scope correctly is vital and should be a team effort – taking on board all expertise.

If the Scope is silent, then the only tests and inspections that the *Contractor* is required to undertake are those required by the applicable law (clause 41.1).

If you suspect a Defect, you can, as the *Supervisor*, instruct the *Contractor* to search for that Defect. However, if there is no Defect, then this will be a compensation event (clause 60.1(10)).

It sounds like you are better off redrafting the Scope and then getting the *Project Manager* to instruct those changes to the Scope (clause 14.3). This, however, will be a compensation event (clause 60.1(1)). It may be more sensible to get the *Project Manager* to notify an early warning and discuss with the *Contractor* the impact of the additional tests and inspections you now require – a view can then be taken.

Ultimately, if the work does not comply with the quality standards within the Scope, then it will be a Defect (clause 11.2(6)).

58. What does 'unnecessary' mean? (clause 41.5)

QUESTION

I am trying to understand what 'unnecessary' means under clause 41.5 of the ECC. I recognise that if a test or inspection by the *Supervisor* causes unnecessary delay then this would be a compensation event (clause 60.1(10)) – but what does this mean?

ANSWER

This is a good question – and its answer could be subjective. Our advice would be to state the time periods within the Scope. Tests and inspections normally fall into three categories: tests and inspections that the *Contractor* undertakes, those undertaken by the *Supervisor* and those witnessed by the *Supervisor*. Why not create a table in the Scope and, against those that involve the *Supervisor*, include a timescale?

59. Charging costs incurred for repeating tests and inspections (clause 41.6)

QUESTION

As the *Client*, we have engaged a consultant *Supervisor* on an hourly rate to manage testing and Defects on our ECC project. We are having several problems with our *Contractor*, and numerous tests need to be repeated because of Defects being found when tests are undertaken. Are we allowed to charge the *Contractor* for this?

ANSWER

Yes – clause 41.6 allows the *Project Manager* to assess the cost to the *Client* in repeating a test or inspection. The *Contractor* pays the amount. While the work is underway, the cost will be deducted from the amount due (second bullet of clause 50.3).

60. Is there a snagging list? (clause 43.2)

QUESTION

Reading clause 43.2 in conjunction with clause 13.7 appears to imply that every single Defect should be notified separately – surely this cannot be right as we have hundreds of Defects on our project?

ANSWER

There is no ambiguity. Clause 43.2 requires that until the *defects date* the *Supervisor* and *Contractor* notify one another of each Defect. The 'notification' needs to be undertaken separately as per clause 13.7, as you say. Each one may have a differing *defect correction period*, depending on how these have been completed in the Contract Data, so it is important that they are separate.

This rigour ensures that we overcome the problems of old – poor and ambiguous communications. Perhaps there will be more communications, but it will be very clear where everybody stands with each Defect. Now that most projects rely on online communications systems, the quality of notifications is much less of an issue. It is useful to agree a communication protocol at the outset of the project, agreeing how this will be managed.

61. Pro forma for latent Defects (clause 43.2)

QUESTION

Is there a latent Defects pro forma?

ANSWER

Firstly, we are not sure about 'latent' Defects – latent to what? Assuming that you are using an ECC contract, then all Defects (note the definition in clause 11.2(6)) found up until the *defects date* are notified by the *Supervisor* to the *Contractor* (clause 43.2). You can use a generic notification form, or you can design your own specifically for this. Strictly speaking, none of these are latent Defects as such. After the *defects date*, any Defect the *Client* finds, which is not listed on the Defects Certificate, may well, in law, be a latent Defect, but you will need to get legal advice on that, and the answer will partly depend on the *law of the contract*. The contract keeps the *Client's* rights regarding these (last sentence of clause 44.3). The law will then decide how and when these should be notified. However, in such a case there is no automatic expectation that the *Contractor* will correct the Defect (as there is before the *defects date*), and you will need to take legal advice as to how to proceed.

62. Constraints in the Scope (clause 43.2)

QUESTION

We are working under an ECC contract with main Option C. The *Contractor* has designed a heating system to provide a room temperature of 10°C. The Scope states this should be 15°C. I, as the *Project Manager*, have therefore advised the *Contractor* that I am disallowing costs under the eighth bullet point of clause 11.2(26), as this is a Defect associated with the *Contractor* not complying with a constraint as stated in the Scope. This has been challenged on the basis that the Scope does not contain a schedule of constraints (i.e. compliance or not with a specification clause is not deemed as a constraint). Do constraints have to be specifically identified in a separate document, highlighted in say a list, or can they just be items included within the Scope?

ANSWER

We assume that the specification is within the Scope. We consider that you were wrong to treat this as a Disallowed Cost, but not for the reasons the *Contractor* suggests. The constraint is on 'how' the *Contractor* Provides the Works, not on 'what' it provides, as is the case here. This only applies to method specifications, not results-based specifications, as this one clearly is.

Such constraints do not have to be in a formal schedule of constraints as is suggested by the *Contractor*, but they do have to be clear and unambiguous. However, that does not apply in your case for the reason given previously. This is, of course, still a Defect that the *Contractor* must correct (clauses 11.2(6) and 44.1).

The *Supervisor* (not *Project Manager*) should therefore notify it under clause 43.2. The cost of correcting this Defect will be paid as Defined Cost, if it is corrected before Completion (seventh bullet of clause 11.2(26)).

63. Dealing with a Defect (clause 44)

QUESTION

On our project, the *Client's* Scope specifies some mesh access flooring to be 45 mm thick. The *Contractor* has installed this at 43 mm thick on an ECC Option C contract. The work is underway – how do we best deal with this?

ANSWER

The use of the 43 mm mesh will be a Defect because it does not comply with the Scope (first bullet of clause 11.2(6)). The *Supervisor* (not *Project Manager*) should therefore notify the *Contractor* of this Defect (clause 43.2). The *Contractor* is then required to correct it under clause 44.1. That must be carried out before Completion, if it will prevent Completion (clause 11.2(2)), otherwise it must be carried out within the *defect correction period* set out in the Contract Data (clause 44.2). If the *Contractor* does not correct this within that period, then the *Project Manager* assesses what it will cost the *Client* to get it corrected by other people and deducts that from the *Contractor* (clause 46.1). As to payment, we note that this is an Option C contract, which shares the risk of Defects between the Parties. The answer will therefore depend on when the correction is carried out. If it is carried out after Completion, then the cost of it will be a Disallowed Cost (fourth main bullet of clause 11.2(26)). Otherwise, the cost forms part of Defined Cost, and is paid. In Option C, the risk of the *Contractor's* errors is, with few exceptions, shared between the Parties. If the *Client* decides that the 2 mm discrepancy is not important, then the Defect could be accepted under clause 45.

64. Dealing with Defects after Completion (clause 44)

QUESTION

We have an ECC project. The project achieved Completion 2 months ago, and we are debating with the maintenance team what kind of warranty periods the ECC provides for one aspect – the air-handling units.

ANSWER

The 'warranty' period in the ECC is effectively the period between Completion and the *defects date*, albeit the contract does not use the term 'warranty'. This is frequently 52 weeks, but it is whatever has been completed in the Contract Data. During this period, there is a dual obligation – the *Supervisor* and *Contractor* need to notify one another of Defects and the *Contractor* is required to correct them within the *defect correction period* (again identified in the Contract Data).

In addition, even when the Defects Certificate has been issued that does not affect your rights for that breach (last sentence of clause 44.3). But this final point will depend on the *law of the contract*.

65. No operational and maintenance manuals! (clause 44)

QUESTION

I undertook *Project Manager* duties on a small ECC contract last year. The *Contractor* has now contacted me for payment of retention money (2.5% outstanding); however, the *Contractor* has not provided the operational and maintenance manuals that were detailed in the Scope. Should I have notified the *Contractor* that this was a Defect for non-compliance with the contract? And although I did not, can I still notify a Defect and withhold the retention money until this information is provided? It is my understanding that I can withhold approval of retention money until the Defects Certificate has been issued.

ANSWER

The *Supervisor* should have notified these omissions as Defects as soon as it was aware of them. Whether it can do so now depends on the *defects date*, which is defined in the Contract Data as being a period after Completion. If the *defects date* has passed, it is now too late to notify a Defect (clause 43.2). In that case, all the *Supervisor* can do now is issue the Defects Certificate, certifying all notified Defects have been corrected (clauses 44.3 and 11.2(7)). The *Project Manager* is then required to issue the final certificate (clause 50.1), which will include the release of all retention (clause X16.2). If the *defects date* has not yet passed, the *Supervisor* (not the *Project Manager*) should immediately notify the *Contractor* that these omissions are a Defect. The *Supervisor* does not need to instruct the *Contractor* to correct them because the contract already does that. The *Contractor* then must correct those Defects within its *defect correction period*, which starts when the *Supervisor* notifies the *Contractor* of the Defects (clause 44.2). The *defect correction period* is set out in the Contract Data. If the *Contractor* does not correct these within this period, then the *Project Manager* assesses what it will cost the *Client* to get these corrected (by other people), and the *Contractor* (clause 46.1) pays that amount. The return of any retention is not explicitly linked to individual Defects and relies on the correct issue of the Defects Certificate for any payment to be due from the *Contractor*.

66. Disputing the Defect (clause 44)

QUESTION

The ECC User Guides state that the term 'Defect' has a restricted definition and that it is intended to include unfinished or omitted parts of the *works*. We currently have an ECC Option A contract, which is in the period between Completion and the *defects date*, and, as the *Client*, we are experiencing issues with the roof of a building constructed by the *Contractor*. The Scope required a slate roof to be provided for the building, with the *Contractor* responsible for the design. Following Completion, we have had dozens of tiles falling from the roof, which we view as being caused by a poor batten design within the roof structure that allows wind to blow into the void and disturb the tile fixings. However, this issue was only discovered after take over, and therefore we are seeking advice as to whether this can be a Defect for notification following Completion. The *Contractor* is contending that it is not responsible for repairing the roof or resolving the underlying cause. If the issue is classed as a Defect and the *Contractor* maintains its position, then we propose to invoke our right under clause 46.1 to assess the cost of having the Defect corrected by other people and to deduct this from the *Contractor*.

ANSWER

This certainly is a Defect. The Scope requires the *Contractor* to design and install a functional roof, which it has clearly failed to do. The fact that this has not happened until after Completion (and/or take over) is irrelevant.

The *Supervisor* (not *Project Manager*) should immediately notify this as a Defect if it has not already done so. If the *Contractor* does not correct the Defect within the *defect correction period*, then the *Project Manager* should assess the cost of getting others to repair it (clause 46.1) and the *Contractor* will pay that amount.

67. Quality control in the ECC (clause 44)

QUESTION

Does the ECC automatically or otherwise give the *Client* a 'right to reject' the work done by the *Contractor* if this does not meet agreed requirements?

ANSWER

The main obligation on the *Contractor* in terms of quality arises in clause 20.1, which states that the *Contractor* Provides the Works in accordance with the Scope. A Defect in clause 11.2(6) is a part of the *works* that is not in accordance with the Scope.

A *Supervisor* is generally appointed by the *Client*. The *Supervisor* and *Contractor* have a reciprocal obligation under clause 43.2 to notify each other as soon as they find a Defect up to the *defects date*.

Clause 44.1 gives the obligation to the *Contractor* to correct the Defect. Clause 44.2 states by when this is to be corrected, and there are other clauses dealing with uncorrected Defects and accepting Defects.

68. Consequential costs of Defects (clause 44)

QUESTION

Testing and Defects are covered under section 4 of the ECC, but is there any provision in the contract to recover or offset the cost for damage caused by the *Contractor* or the recovery of consequential loss by the *Client*?

ANSWER

The ECC sets a period during which the Parties have dual obligations if a Defect is found. In the ECC, that period ends at what the contract calls the *defects date*. Up until then, the *Contractor* is required to correct any Defects that are found (clause 44.2), and the *Client* is obliged to provide access so that the *Contractor* can do so (clause 44.4). Neither can 'charge' the other for this, and each covers its own costs. So, the *Client* is unable to recover consequential costs. Latent Defects found after the *defects date* will be dealt with as breaches of contract, and the dual obligation no longer applies. Damage caused by the *Contractor* when correcting a Defect will be the *Contractor's* liability (sixth main bullet of clauses 80.1 and 81.1). The *Contractor* is required to insure against this (clause 83.2).

69. Defects arising after the Defects Certificate is issued (clause 44.3)

QUESTION

I have the job of closing out schemes that were completed several years ago but, for various reasons, are still in an aftercare and landscaping period, although the Defects Certificate has been issued. The majority of the schemes are ECC Option C and signed by deed, which have a 12-year latent Defects period provided for by law in the country where we are working. My understanding of a latent Defect is something that is not obvious; for example, it could be a geotechnical Defect. If we suspect a Defect has arisen after the Defects Certificate has been issued, but within the 12-year period, what should we do?

ANSWER

The contract says nothing about Defects after the issue of the Defects Certificate. The issue of the Defects Certificate does not relieve the *Contractor* of all its obligations with regards to Defects (clause 44.3). You will need to get legal advice on the *Contractor* remaining liable for breaches, as you will have to show that whatever went wrong was caused by the breach of the contract by the *Contractor* and work within the Limitation Act 1980 if your projects are in England, Wales or Northern Ireland. The contract is, quite deliberately, silent on the matter, and has no procedures to deal with it, to allow whatever law the contract is under to do so instead. Different rules will apply in other legal jurisdictions.

70. Accepting a Defect (clause 45)

QUESTION

Our *Contractor* is proposing that a Defect is not corrected on our project and suggested that it could offer a cost saving to us. Is this permissible under the ECC and how should it be valued?

ANSWER

Yes, the *Contractor* or the *Project Manager* may propose to the other that the Scope is changed so that a Defect does not have to be corrected (clause 45.1).

The proposal from the *Contractor* may result in an earlier Completion Date and a reduction in Prices. There is no methodology as such stated in the ECC (like clause 63.1 requires for the assessment of compensation events); however, the figure needs to be sensible and agreed between the *Project Manager* and *Contractor*.

Also, when considering the monetary assessment, the potential cost of not doing the work should be considered (if it is simply an omission, i.e. the *Contractor* has failed to do something). In addition, life cycle cost issues need to be considered. Under clause 45.2, once accepted the Scope is changed accordingly – this means that the *Contractor* is no longer responsible for the Defect. If the Defect being accepted now incurs additional life cycle cost, then this should also be a factor in the reduced Prices (i.e. the installation of a less efficient pump may result in a higher utility cost to run).

71. Not getting access to correct Defects (clause 46)

QUESTION

We are under the ECC Option C form of contract. As a *Contractor* we have been notified of a Defect, although we have not yet achieved Completion. The correction of Defects under these circumstances would be a Defined Cost and therefore part of the Price for Work Done to Date. However, what stops the *Project Manager* from deliberately not granting access for the correction of a Defect – thus ensuring that the *Contractor* cannot correct the Defect (and include it in the Price for Work Done to Date) – and just use clause 46.2 to charge the *Contractor* the cost of correcting the Defect? This would appear to prejudice the *Contractor's* position in being paid (although this will affect the *Contractor's* share) and artificially create a situation in which the *Contractor* would have to pay for the correction of all Defects.

ANSWER

We think you may be confusing two different matters: whether you get paid to carry out work, and what your obligations are to correct Defects. If, because of the definition of Completion, the Defect requires correction before Completion can be certified, then the *Project Manager* cannot certify Completion until it has been corrected.

Therefore, we cannot see how this scenario can happen, because it is not a Disallowed Cost until after Completion has been certified in accordance with the contract. In addition, the *Client* is required to allow you access to and use of the Site (clause 33.1) up until take over, and if it does not that will be a compensation event (clause 60.1(2)). You have such access and use until take over, and just because the *works* are taken over does not mean that Completion has been achieved. Clause 46.2 only applies once Completion is certified. After that you are required to correct Defects (clause 44.2), and under Option C the cost of that will be a Disallowed Cost.

So, after Completion, Defects are at your cost anyway, with or without clause 46.2. You need to be aware that the wording of clause 46.2 is deliberately different from that in clause 46.1. If the *Client* cannot provide access for you to correct a Defect, the value of the deduction is not the cost of getting somebody else to correct the Defect; it is what it would have cost you to correct. This means that you are in no worse a position than if you had carried out the correction yourself.

72. Whether the *Contractor* comes back or not (clause 46)

QUESTION

I am trying to understand the difference between clauses 46.1 and 46.2 of the ECC. They do not make sense to me – can you help?

ANSWER

Yes, these outline two scenarios that relate to uncorrected Defects. Scenario 1 (clause 46.1) is when the *Contractor* was given access to correct the Defect by the *Client,* but it failed to rectify it within the *defect correction period.* In this instance, the monetary value is assessed as the cost incurred by the *Client* of having the Defect corrected by other people – this will typically be a higher cost than clause 46.2. Whether the Defect is corrected by other people is up to the *Client*, but the monetary value is assessed in this manner.

Scenario 2 (clause 46.2) is when the *Contractor* was not given access by the *Client* – perhaps because the building was operational, and it would have been impossible/unsafe to rectify the Defect. In this instance, the lesser value would be assessed – this cost was incurred by the original *Contractor* in rectifying the Defect. In theory, this should be a lesser cost than clause 46.1. Again, it is then the *Client's* decision as to whether the Defect is corrected.

Both scenarios provide a sum of money for the *Client* to then deal with the problem.

73. Defects after the Defects Certificate is issued (clause 43.2)

QUESTION

What happens with Defects after the Defects Certificate has been issued? The *Project Manager* on our job keeps on notifying Defects.

ANSWER

It is the obligation of the *Contractor* and *Supervisor* to notify each other of Defects as soon as they become aware of them, not the *Project Manager* (clause 43.2). And this obligation only runs until the *defects date*. Should defective work be discovered after this point, the *Client* will need to take legal advice as to general law remedies available.

FURTHER THOUGHTS AND TOP TIPS

- The Scope should clearly define the quality standards.
- Ensure that the *defect correction period(s)* are appropriate and thought through.
- Ensure that the Client manages Defects post-Completion with its resources. Often, they are not managed as the contract requires.
- If tests or inspections are undertaken by the Supervisor, include their timescales and durations.
- Do not be afraid to notify Defects. It is what the contract requires. Following the contract is professional and provides a clear audit trail.

Robert Alan Gerrard and Patrick Waterhouse
ISBN 978-1-83708-959-8
https://doi.org/10.1108/978-1-83708-956-720252005
Emerald Publishing Limited:

Chapter 5
Payment

74. Making applications for payment (clause 50.1)

QUESTION

In an Engineering and Construction Contract (ECC) Option C contract, once the Completion certificate has been issued, is there a set period in which the *Contractor* can still submit applications for payment? Once the Completion certificate has been issued, does the *Contractor* have 4 weeks to submit its application and after this date, if it is still working on the final application, can it still submit an application or is this then at its own cost?

ANSWER

The *Contractor* is obliged to submit an application for payment before each assessment date to the *Project Manager* (clause 50.2). The *Project Manager* is obliged to assess the amount due at each assessment date. If the *Contractor* has made an application before the assessment date, the *Project Manager* must consider it (clause 50.2), but if the *Contractor* has not, the *Project Manager* still must make its own assessment (clause 50.4). Assessment dates continue at the maximum of the assessment intervals from the *starting date* until 4 weeks after the issue of the Defects Certificate by the *Supervisor* or the *Project Manager* issues a termination certificate (clause 50.1). So, assessments are made by the *Project Manager* after Completion and payments made, to the extent that the Price for Work Done to Date has changed. And that is especially so with Option C, under which you cannot make the final payment of the *Contractor's* share until you have finalised the Price for Work Done to Date and the final total of the Prices (clause 54.4).

75. Calculating the final account (clause 50.1)

QUESTION

We are engaged as the *Project Manager* on an ECC Option C project in which the *Contractor* has failed to provide its final account. Completion was achieved a while ago. Do we have to provide any notice before we make a *Project Manager's* assessment for the final account?

ANSWER

The obligation is on the *Project Manager* to make an assessment of the final amount due and to certify a final payment. This will be no later than 4 weeks after the *Supervisor* issues the Defects Certificate unless a termination certificate is issued (clause 53.1).

So, you should already be assessing what is due and certifying it. In Option C, it is important you set up a system whereby the *Contractor* shows the *Project Manager* its Defined Cost each month and the *Project Manager* regularly audits this. The *Project Manager* will fail in its duties if it certifies payment based on 'say' or unsubstantiated figures (see the first bullet of the definition of Disallowed Cost in clause 11.2(26)). This exercise should not be left until the end of the contract, as it can lead to all sorts of problems, arguing retrospectively about what is and is not Defined Cost and Disallowed Cost.

Therefore, by the time you get to Completion under Option C, all the hard work should be done, and the *Project Manager* should have a good idea of what the total Defined Cost is. And until such time as the *Contractor* justifies any other costs, they are not due because they will be a Disallowed Cost until they are justified.

As to the payment of share (one way or other), the main tranche of that should have been certified by the *Project Manager* at the assessment immediately following Completion and should be based on its forecasts of the final Price for Work Done to Date and the final total of the Prices (clause 54.3). The final payment of the share is made as soon as the final figures for the Price for Work Done to Date and total of the Prices are known (clause 54.4).

The final amount due must be certified within 4 weeks of the *Supervisor* issuing the Defects Certificate (clause 11.2(7)), which in turn must be issued by the later of the *defects date* and the end of the last *defect correction period*. So, the *Project Manager* should make its final assessment of the amount due by then.

76. Assessing the amount due (clause 50.3)

QUESTION

We have an ECC Option A contract that incorporates secondary Option X7 delay damages and Option X16 retention. When assessing the amount due, should the *Project Manager* deduct retention before or after applying delay damages?

ANSWER

The amount due at each assessment date is detailed in the three bullets of clause 50.3. The clause states that it is the Price for Work Done to Date, plus other amounts to be paid to the *Contractor*, less other amounts to be paid by the *Contractor*. Clause X16.1 makes it clear that you deduct retention from the Price for Work Done to Date less the *retention free amount*. It is not deducted from the other amounts in the other two bullets of clause 50.3. You therefore calculate the Price for Work Done to Date, then deduct the retention. After that, the delay damages are deducted under the third bullet of clause 50.3. The *Project Manager* therefore deducts retention before applying the delay damages.

77. Retaining money due to non-compliant programmes (clause 50.3)

QUESTION

We have a query on an ECC Option C contract related to the Accepted Programme. At the outset of the contract, we had a programme identified in Contract Data part two. This was further developed by the *Contractor* to provide more detailed information – that is, it went to a lower level to identify the deliverables associated with the associated tasks. This programme was then used by all Parties as the managing tool and was updated and issued on a monthly basis as required under clauses 32.1 and 32.2.

The project has now been going for some 2 years, and in this period the programme has only been formally rejected twice. There has never been a formal acceptance of any programme. While the programme issued under clause 32 and the one in Contract Data part two have been used by both Parties, they have never strictly complied with the requirements of clause 31.2.

The *Client* is now proposing to retain one quarter of the Price for Work Done to Date in accordance with clause 50.5. While we appreciate that this should be retained only in relation to the first programme, can this still be applied, as the programme is not in accordance with clause 30.1 and has never been formally accepted?

ANSWER

The simple answer is no; the *Client* cannot do that. For a start, only the *Project Manager* has that authority. Clause 50.5, which deals with the possible retaining of one quarter of the Price for Work Done to Date, is very clear. It starts with the words: 'If no programme is identified in the Contract Data'. As there is a programme identified in the Contract Data, clause 50.5 simply does not apply at all.

It is quite surprising that the *Client* is even suggesting doing this. For the last 2 years the *Contractor* has done what the contract requires of it, in that it has produced and submitted programmes for acceptance at the interval stated in the contract. In contrast, during that time the *Project Manager* (which works on behalf of the *Client*) has failed almost completely to do what it is supposed to do under the contract. The *Project Manager* is supposed either to accept or not accept each programme within 2 weeks of issue (clause 31.3). Instead, it has, for the most part, ignored the submissions, which are breaches of contract on its part that have resulted in a whole series of potential compensation events under clause 60.1(6). To suggest that the *Contractor* should now be punished for some theoretical wrong at this late stage does not seem to be sustainable.

In addition, that failure now puts the *Client* in a weak position when it comes to disputes as to time. To suggest later that the programmes issued by the *Contractor* for acceptance were somehow 'wrong' will be very difficult. With two exceptions, the *Project Manager* has never declined to accept them, as the contract requires it does if they are wrong, and then those supposedly 'wrong' programmes were used by both sides to manage the *works*.

78. Application for payment (clause 50.4)

QUESTION

Is there an obligation for the *Contractor* to submit an application for payment in the ECC? If so, what happens if it does not – does the *Project Manager* still need to make an assessment?

ANSWER

Yes, there is an express obligation for the *Contractor* to submit an application for payment. Under clause 50.4 the *Contractor* is required to make an application for payment. If it fails to do so, the amount due is the lesser of

- the amount that the *Project Manager* assesses as due at the assessment date
- the amount due at the previous assessment date.

The first bullet addresses any potential negative adjustment in payment – to take account of the delay damages, for example.

The ECC therefore contains a direct obligation and incentive for the *Contractor* to submit an application for payment. Without such there will be no further positive adjustment to payment.

79. No compliant programme is ever submitted (clause 50.5)

QUESTION

Clause 50.5 of the ECC states

> If no programme is identified in the Contract Data, one quarter of the Price for Work Done to Date is retained in assessments of the amount due until the *Contractor* has submitted a first programme to the *Project Manager* for acceptance showing the information which the contract requires.

My question is, what happens to the retained money should the *Contractor* fail to submit a compliant programme (or indeed, a programme of any kind) throughout the contract period? In anticipation of the possible reply that a *Project Manager* should not allow the contract to continue without an Accepted Programme, in our case the *Client* is unlikely to want termination on the grounds that the *Contractor* has not provided a programme – particularly when the *works* are seen to be progressing in an apparently organised manner.

ANSWER

Without an Accepted Programme the contract cannot be properly managed. Several compensation events rely on an Accepted Programme (clauses 60.1(2), 60.1(3) and 60.1(5)). With no Accepted Programme, how can you make clause 63.5 work in respect of moving the Completion Date?

The lack of a programme is indicative of one of two things. Either the *Contractor* is disorganised, despite appearances to the contrary, and it has no way of knowing whether it is going to finish by the Completion Date. Or the *Contractor* has a programme that it does not want to show you so that it can do things the 'old-fashioned' way – that is, make one up afterwards, once it knows where the delays have occurred, to ensure that the critical path runs through all those delays that you caused, and none that the *Contractor* caused. Neither is good news for your project, and both should be avoided at all costs.

On the other hand, the *Project Manager* should not use the list in clause 31.2 as a 'rule book'. As long as it is satisfied that the programme is achievable and has all the information it needs to administer the contract on behalf of the *Client* (such as where the critical path is and what resources have been assumed), then it should accept it, and release the one quarter of the Price for Work Done to Date being retained. Without an Accepted Programme, the assessment of all compensation events is taken out of the hands of the *Contractor* altogether. The *Project Manager* is required not to accept any compensation event quotations and make its own assessment of the compensation event (third and fourth bullets of clause 64.1). And that assessment must be made using the *Project Manager's* assessment of the programme for the remaining works (clause 64.2).

As for when the one quarter is released, the contract is silent, because it did not envisage that any *Contractor* would allow this situation to continue for any length of time, given that cash flow is the lifeblood of any *Contractor's* business. Because there is no date for release other than the production of a programme, there is one argument that says that the *Client* keeps it, although it may not get a sympathetic ear from an *Adjudicator* or a court if the *Contractor* were to dispute that. The money is retained by the *Client*; it does not have ownership of it so it is likely that the amount would be returned in the final assessment.

We would strongly suggest that the *Project Manager* actively manages this situation. Without a programme there is a real risk that Completion will be delayed. The *Project Manager* should notify an early warning to the *Contractor*, ideally attending an early warning meeting as soon as possible with senior members of the *Contractor's* organisation to sort this problem out.

80. Retaining monies when a programme is not accepted (clause 50.5)

QUESTION

On an ECC contract we have an initial programme; however, we have since had compensation events for which we believe we have not received adequate particulars covering additional cost or programme alterations. The latest revised programme shows a period of time risk allowance having been used up, and the effects of the compensation events are still unknown. We consider this programme is not acceptable. Can the *Project Manager* therefore deduct a quarter of the monies owed to the *Contractor* for this, or does this provision apply to the first programme only?

ANSWER

If the initial programme was supplied and referred to in Contract Data part two under the heading 'The programme identified in the Contract Data is …', then you cannot apply the quarter of the Price for Work Done to Date retention. This is because clause 50.5 makes it clear that it only applies if no programme is identified in the contract data. However, if

- there was a programme issued with the tender that was not referred to in the Contract Data and
- the programme you now have is the first one issued for acceptance since the contract came into existence, and
- the programme does not show all the matters listed in clause 31.2

then you must apply the retention.

There are, though, two other things you need to be aware of. First, this retention does not depend on the *Project Manager* accepting the programme. If it has been submitted and shows all the information it should, then the retention is not taken even if the *Project Manager* thinks it is wrong for some other reason. Secondly, this is only a retention, not a deduction, and eventually it will have to be paid back.

81. Pay less notices and delay damages in the Engineering and Construction Short Contract (ECSC) (clause 50.3)

QUESTION

Under the ECSC, can delay damages be deducted without any need for a pay less notice to be issued?

ANSWER

We assume that this question refers to a contract in the UK subject to the legislation for construction contracts. In which case the additional conditions of the ECSC apply. Looking at clause 50.3, the deduction of delay damages is part of the amount due (third bullet). However, this contract works on the trigger for payment being the *Contractor's* application (clauses 50.2 and 50.4) and the amount stated in the *Contractor*'s application being the notified sum. Therefore, if the application does not contain the delay damages, you will dispute the notified sum and a pay less notice must be issued. See additional condition 1.4, which replaces clause 50.5.

We would therefore strongly recommend that you always provide a notice of what you are going to pay if it is not what the *Contractor* has applied for, with details of what deductions you are making and why.

82. Error in the ECSC Price List (clause 50.3)

QUESTION

I have a query regarding the ECSC. If a tenderer makes an arithmetical error in the Price List, should this error be corrected when the tenders are opened, and the revised figure inserted in the *Contractor's* offer? Or can the tendered sum not be changed?

ANSWER

It is entirely up to you how you deal with this at the tender stage. You have no contract at the tender stage so there are no rules that are relevant within the contract (the same as other contracts). It will also depend on what the *Client's* policy is on the matter, and what the instructions to tenderers said you would do in the circumstances. Some clients adjust the total, other clients ask the tenderer if it wishes to stand by its total, and, if so, what adjustments it wants to make in the prices in the Price List. Some clients would leave the figures as they are but use the correct total to assess the tenders. There is no wrong or right approach. As to what happens if you accept the tender with this error in it, the contract is clear. The *Contractor* is paid the Price for Work Done to Date (clause 50.3). The definition of Price for Work Done to Date is in clause 11.2(12), and you will see it is based solely on the figures in the Price List. So, the *Contractor* will get paid the correct total, not the incorrect figure it added it up to be at the tender stage. There may also be procurement regulations for you to consider depending on the law in the jurisdiction in which you are based.

83. Difference in payment periods between the ECC and ECSC (clause 51)

QUESTION

Clause 51.2 in the ECC states that the *Client* must pay the *Contractor* within 3 weeks of the assessment date. Clause 51.1 in the ECSC states that the *Client* must pay the *Contractor* within 7 weeks of the assessment date. Why are these so different?

ANSWER

We disagree with your opinion that the ECSC provides for a 7-week payment provision. It provides for 3 weeks after the next assessment date. Clause 50.1 states that the *Contractor* has to make an application before each assessment date. So, if the *Contractor* gets its assessment of the amount due to the *Client* by the day before the assessment date, it will be paid 3 weeks later. However, if it misses that deadline and does not get it in until the day after, then it will likely have to wait until the following month to receive any monies (clause 50.4). So, if the *Contractor* does what clause 50.1 of the contract requires, it will be paid in 3 weeks. If it does not do what is required, it will probably have to wait another month. So, there is no difference between the ECC and ECSC, if the *Contractor* follows the requirements of the contract.

84. Can the *Contractor* owe money to the *Client*? (clause 51.2)

QUESTION

In the ECC, should the *Contractor* be required to pay damages from one of the secondary Option clauses or amounts for uncorrected Defects, and is the *Project Manager* required to certify as per the requirements of the core clauses in section 5 by reducing the amount due by the relevant amount, and therefore creating a situation in which the *Contractor* would owe the *Client* that amount due?

To add some context to this, we have a situation in which damages and uncorrected Defects may create a situation where the *Contractor* must pay the *Client*. The *Project Manager* has decided that it can just demand payment almost immediately and without any form of certification; we state that the provisions and payment times are dictated by the core clauses in section 5.

ANSWER

The only way that payment becomes due from one Party to the other is through the operation of the assessment and certification process set out in section 5 ('Payment') of the contract. The *Project Manager* is required to issue a certificate for the amount due in accordance with clause 50. The timing is set out in clause 50.1, and runs all the way until the issue of the Defects Certificate (or the *Project Manager* issues a termination certificate). The amount due is set out in clause 50.3. The sums that you are referring to would fall under the third bullet of clause 50.3 'less amounts to be paid by or retained from the *Contractor*'. The *Project Manager* then must certify in accordance with clause 51.1.

If that certificate reduces the amount due from that previously paid by the *Client*, then clause 51.1 requires the *Contractor* to make the payment within the timescales set out in clause 51.2. If you do not pay by that date, interest becomes due to the *Client* at the rate in the contract, just as it would become due to you if the *Client* does not pay on time.

When deductions can be made depends on what they are being made for. The earliest that the *defect correction period* can run from is Completion (clause 44.2), and so there can be no deductions for your failure to correct Defects until after Completion. The same applies to delay damages. If the *Project Manager* wants to recover these from you, then it must assess and include them in the next assessment that it is due to make under clause 50.1. That will usually be monthly. The *Project Manager* then follows the processes in clauses 50 and 51, and you pay in accordance with clauses 51.1 and 51.2. By the way, if you do not intend to pay, you must issue a pay less notice in accordance with clause Y2.3 (or whatever other part of the contract complies with the Housing Grants, Construction and Regeneration Act 1996). The requirement for such a notice falls on whichever Party is not intending to pay, not just the *Client*.

85. Payment in the ECSC (clause 51.2)

QUESTION

We are currently working under an ECSC contract. The *Client* has stated that, under clause 51.1, it can take up to a further 3 weeks to pay after the next assessment date, which would be, say, 4 weeks later. This would be a total of 7 weeks from the original assessment date. Our normal terms under the ECC, for example, are that the *Project Manager* certifies within a week of the assessment date, and that the certified payment is made within 3 weeks of the assessment date (clause 51.2). The ECSC User Guides state that payment relates to an assessment date, which we always assumed was the original date, but did not think it would be the next one. Can you advise?

ANSWER

It seems that the *Client* has misunderstood the contract. The ECSC uses the principle of an assessment date, which is set out in the Contract Data (on p. 2). It could be 'the last day of the month' or the '28th of the month' or 'the last Friday of each month', or any of those sorts of descriptions. You will need to check that. You are then required to make an application for payment in each period (clause 50.3) before each assessment date (clause 50.1). You will then be paid 3 weeks after the next assessment date following receipt by the *Client* of your application (clause 51.1).

Let us say that your assessment date is the 28th of each month, and we are looking at April. If the *Client* receives your application by 27 April, then the next assessment date is 28 April, and you will therefore be paid 3 weeks after 28 April. If the *Client* does not receive it until 29 April, then the next assessment date is 29 May, and you will be paid 3 weeks after 29 May. That means that if the *Client* receives your application before the assessment date, you will be paid within 3 weeks of that assessment date. If you miss that assessment date (even by a day), you must wait until the following month (clause 50.4).

86. Can we add corporation tax to the Fee? (clause 52.1)

QUESTION

Under ECC Option E can corporation tax be included in the *fee percentage* after award?

ANSWER

The ECC does not work like that. The contract defines what you will be paid for as Defined Cost (clause 11.2(24)), and then states that any other cost you incur or any other monies you want to recover have to be included in your Fee (clause 52.1). Corporation tax is clearly not included within the Schedule of Cost Components (SCC) and is therefore not paid as a part of Defined Cost.

It is entirely up to you what *fee percentage* you insert in the contract and what allowances you want to include within your *fee percentage* and at what level. You can add whatever you want to the *fee percentage*, and once it is set in the contract that is the end of the matter, and the *Client* cannot audit it in any way.

The ECC User Guides point out what is not included within Defined Cost and what you need to think about when assessing what *fee percentage* to tender. The guidance is not saying that you should or should not allow for any of the listed items in your Fee, because that is a commercial decision only you can make. However, if you do not allow for them, you will not get paid for them anywhere else.

The addition of corporation tax would seem odd to us, whatever the contract says.

87. What to include in the Fee? (clause 52.1)

QUESTION

I am a *Contractor's* estimator pricing an ECC Option C contract and cannot see where I price nor where we would be reimbursed for the likes of catering supplies, bidding costs, head office overheads and profit.

ANSWER

Simply, Defined Cost is the cost of the components in the SCC less Disallowed Cost (clause 11.2(26)). So, as an estimator you will need to research items of cost that you can recover under the SCC.

Any costs not defined will need to be included in the Fee: 'All the *Contractor's* costs which are not included in Defined Cost are treated as included in the Fee' (clause 52.1) – that is, they are 'deemed' to be in the Fee.

The items you refer to, among others, are certainly not part of Defined Cost, so these will need to be included in your Fee.

88. Passing on discounts (clause 52.1)

QUESTION

We are the *Contractor* on an ECC Option B contract. A compensation event has arisen that includes some Subcontractor works (the Subcontractor is already on Site). We have previously agreed a 2.5% main *Contractor's* discount with the Subcontractor for prompt payment. Are we contractually required to pass this discount on to the *Client*?

ANSWER

The simple answer is yes (clause 52.1: 'with deductions for all discounts, rebates and taxes which can be recovered'). However, when assessing compensation events in Option B, you must follow the provisions of the Short Schedule of Cost Components (SSCC) (see the definition of Defined Cost in clause 11.2(23)). Also, the SSCC states that the Defined Cost here is the amount you pay to Subcontractors, and that would include the 2.5% discount.

89. The final assessment (clause 53)

QUESTION

I am a fencing *Subcontractor* on an Option B Engineering and Construction Subcontract (ECS). The *Contractor* produced the *bill of quantities* which became the Bill of Quantities. Over the course of the contract, the *Subcontractor* has submitted the remeasure in the form of a spreadsheet workbook based on the Bill of Quantities. Despite numerous invitations the *Contractor's* surveyor has declined to undertake any joint Site measures. It is now 6 weeks after the Defects Certificate has been issued and the *Contractor* has failed to make an assessment of the final amount due. We then issued our own assessment, which is the spreadsheet workbook, as clause 53.2 of the subcontract. The *Contractor* states that the spreadsheet workbook is not evidence of the remeasure and that we need to prove any changes to the Bill of Quantities. The Price for Work Done to Date is reduced by £500 000 and a negative payment certificate is issued. What should we do?

ANSWER

It appears that the *Contractor* does not agree with your assessment so it cannot be said that you have reached the final payment stage that clause 53.2 anticipates. You have two choices really. The practical one is to quickly try and further evidence your measurements to support your assessment if possible. While that is not strictly provided in the contract, it may save time and money in the dispute resolution process.

Otherwise, you are left to pursue your rights under clause 53.3 depending on whether Option W1, W2 or W3 is incorporated.

90. The *Contractor's* application for payment (clause 50)

QUESTION

In the ECC using Option A, should the *Contractor's* application for payment reflect work completed up to or including the assessment date?

ANSWER

The amount due is decided at the assessment date. See clauses 50.1 and 50.2. The application for payment must also be submitted before each assessment date but it is not completely clear whether it could include work done on the assessment date itself. In Option C, clause 11.2(31) clearly refers to Defined Cost that the *Project Manager* forecasts will have been paid by the *Contractor* before the next assessment date plus the Fee. We think it is reasonable to apply similar principles to other main Options and suggest the application for payment should be for work completed up to and including the day before the assessment date, for Option A contracts.

91. Can the *Client* keep the retained quarter of the Price for Work Done to Date? (clause 50.5)

QUESTION

If the *Contractor* never submits a first programme to the *Project Manager* for acceptance, showing the information which the contract requires, then under clause 50.5, can the *Client* keep the one quarter of the Price for Work Done to Date, or should it be returned to the *Contractor* as a first programme is no longer needed?

While this may be seen as never happening, take a 3-month high value ECC contract where the *Contractor* never issued any programme, does the *Client* get a 25% discount?

ANSWER

We believe the intention of the drafting is of course that this scenario never arises but struggle to see how a correct deduction just because a programme submitted does not comply with one small aspect of the programme requirements could ever lead to the *Client* correctly keeping one quarter of the Price for Work Done to Date.

That said, there is no obvious clause that says as such. While we cannot give legal advice, such a punitive deduction on a high value contract must surely fail the test of fairness. The advice is, just get a programme in that complies with clause 31.2.

It really should not be a big problem, unless serious cash flow issues arise as a result of the 25% deduction.

92. What is a market rate? (clause 52.1)

QUESTION

What is a market rate in the ECC?

ANSWER

Clause 52.1 gives some rules to help answer this question, but it uses slightly different language. It talks about 'open market or competitively tendered prices with deductions for all discounts, rebates and taxes which can be recovered'. So, for those items of Defined Cost subject to this rule, the *Contractor* will need to ensure it can justify this to the *Project Manager* should it be necessary. The *Contractor* may go to the open market to establish such prices, or it may have or seek competitively tendered prices, on this or other jobs it has on its books. A number of local or national suppliers may already have short-, medium- or long-term agreements in place with the *Contractor* that can be used to help justify such Defined Cost lines up with this clause. Remember always to feed in the discounts, rebates and taxes as referred to. Ask questions on this to assure yourself of the correct cost here.

FURTHER THOUGHTS AND TOP TIPS

- Encourage the *Contractor* to make sure that it does not fall foul of clause 50.5 by submitting a first programme showing the information that the contract requires.
- The *Project Manager* should be aware that it must determine the amount due at each assessment date.
- When interest is due, this does not have to be asked for by the receiving Party – it should be included in the next assessment.

Robert Alan Gerrard and Patrick Waterhouse
ISBN 978-1-83708-959-8
https://doi.org/10.1108/978-1-83708-956-720252006

Chapter 6
Compensation events

93. Omitting work under Engineering and Construction Contract (ECC) Option A (clause 60.1(1))

QUESTION

If we omit work from the Scope under ECC Option A, does this result in a negative compensation event?

ANSWER

Any change to the Scope (under clause 14.3) is a compensation event (except for the two bullets under clause 60.1(1)). If you omit work, then the forecast cost saving will be assessed in accordance with clause 63.1 and the Prices will be reduced.

94. Inconsistencies in the Scope (clause 60.1(1))

QUESTION

We have just entered an ECC Option C contract and immediately encountered a problem with the *Project Manager*. The Scope prepared by the *Client* (part 1) requires that the heating system is on two zones whereas our price and our Scope (part 2) states one zone. From our point of view, what we have offered has been accepted and therefore that is what we believe to be the basis of the contract. The *Project Manager* disagrees. Can you confirm our understanding?

ANSWER

Unfortunately, your understanding does not accord with the contract. Essentially, what we have is an inconsistency between the Scope prepared by the *Client* and the Scope for its design prepared by the *Contractor*. This inconsistency needs to be notified under clause 17.1. The *Project Manager* will then decide which is preferred. A change to the Scope is then instructed by the *Project Manager* under clause 14.3. The answer is in the second bullet of clause 60.1(1) – this states that a change to the Scope prepared by the *Contractor* (your part 2), which is required in order to comply with Scope prepared by the *Client* (your part 1), is not a compensation event. So, the *Project Manager* is right, you will need to install two zones and you will not be entitled to a change to the Prices or a delay to the Completion Date. The cost, however, will be a shared risk under Option C, dependent on the share ranges.

95. Why is accepting a Defect not a compensation event? (clause 60.1(1))

QUESTION

I do not understand the first bullet under clause 60.1(1) of the ECC – why is acceptance of a Defect not a compensation event?

ANSWER

A Defect (as defined in clause 11.2(6)) may be accepted under clauses 45.1 and 45.2. It is not a compensation event as it does not follow the assessment methodology defined under clause 63.1 and other clauses.

A quotation for reduced Prices or an earlier Completion Date, or both, is submitted to the *Project Manager* for acceptance. This is effectively a commercial offer, and, from the *Project Manager's* and *Client's* viewpoints, needs to address future maintenance costs. For this reason, it is not subject to the same assessment methodology as a compensation event.

96. Are compensation events claims? (clause 60.1)

QUESTION

This may seem like a basic question but, being new to the ECC, what are compensation events? Are they claims?

ANSWER

This is a good question – compensation events are listed in the contract and, should they occur, the *Contractor* can be compensated potentially in terms of both time and cost. Within any contract we need to define the limit of risk. The ECC defines 21 (in the core clauses) events that are defined as compensation events. Additional compensation events are also found in the main and secondary Options.

Compensation events broadly fall into three main categories: we have changed our mind, we have not done something we said we would and, finally, a risk has occurred over and above a reasonable risk limit of an experienced contractor.

The whole premise of compensation events is that the *Contractor* should be no better or worse off after the compensation event. You will see within the assessment clauses that both time and cost are assessed. The other key difference between the ECC and many other standard form contracts is that compensation events are assessed and what is termed 'implemented' within defined timescales.

97. Stopping the work (clause 60.1(4))

QUESTION

We are working on a petrochemical site and are under Option A of the ECC. The *Project Manager*, due to other projects on the Site, has on several occasions told us to stop work and on other occasions prevented us from doing work. There was nothing mentioned in the Scope about this – and we have incurred cost as a result. Is this a compensation event?

ANSWER

Firstly, it is important to note that you need to obey these instructions (clause 27.3), which it sounds like you have: the *Project Manager* has the authority to give such instructions. And you are right – this is a compensation event under clause 60.1(4). It should be notified by yourselves under clause 61.3 (if not already done by the *Project Manager*). This will start the compensation event process and enable the consequences of the disruption to be assessed.

98. Dealing with Others (clause 60.1(5))

QUESTION

We are delivering a complex engineering project under Option C of the ECC. The *Client* will be involved in reviewing our design as a *Contractor*. The work also encroaches on a railway line, and so will involve the rail operator. We also have potential archaeological issues and utility connections. How do we define the interaction of all these key players?

ANSWER

The answer to this is that it should all be in the programme! The programme is at the 'heart' of the ECC. Clause 31.2 requires the *Contractor* to chart (fourth bullet) the order and timing of the *Client* and Others. The Others are defined (clause 11.2(12)), and basically cover organisations that are not employed by the *Contractor* – that is, they are the *Client's* responsibility.

Failure of the *Client* or Others to work within the times shown on the Accepted Programme, failure to work within the conditions stated in the Scope or the carrying out of additional work would all fall under the realm of a compensation event (clause 60.1(5)). It is important that the programme is managed collaboratively and that the interaction of all these organisations is clearly defined. A suggestion would be to meet with the *Project Manager* and discuss the programme before submission – this will ensure it is accurate and agreed.

99. Late decisions (clause 60.1(6))

QUESTION

I am a *Project Manager* on an ECC Option A contract. I am aware of the need to make decisions within defined timescales but the *Client* I am working for is notorious for not delegating and taking too long to make decisions. I am delivering a board presentation to the *Client* team and want to highlight the key risks – can you advise what the impact of late decisions would be?

ANSWER

You are right that the ECC requires timely decision-making – this is fundamental to any project management approach. Ultimately, failure to respond to a communication is a compensation event (clause 60.1(6)). Let us look at three examples. Failure to respond to a design submission will stop the *works* – the *Contractor* does not proceed with the relevant work (clause 21.2). Failure to respond to communications asking for the acceptance of proposed Subcontractors will prevent the orders being placed – again, having a material impact. However, failure to respond to the programme (clause 31.3) does not 'stop the project' – the *Contractor* would still be obliged to continue. In all these instances, however, the *Project Manager* should respond within the period required by the contract. This is a key message to get across to your board. The ECC assumes that full authority is placed with the *Project Manager*.

100. Objects of interest (clause 60.1(7))

QUESTION

We have a situation on our project in which we found an object of historical interest, but it was in a corner of the Site that did not affect the *Contractor*. So, in principle, it is a compensation event but there was no time or cost impact. How do we close this off? As the *Project Manager*, I do not want to leave things open ended for someone else to deal with later or for new people having different interpretations.

ANSWER

This will be a compensation event if you instructed the *Contractor* how to deal with the object (73.1/60.1(7)). Merely finding the object is not a compensation event. What you need to do is notify the compensation event under clause 61.1 and instruct the *Contractor* to submit a quotation under clause 61.2. This gives the *Contractor* the opportunity to assess the time and cost impact. Subject to this coming back as no time/cost impact, this will be implemented as such (first bullet of clause 66.1), and this is not then revised (clause 66.3). You are right to close things off, as instructions from the *Project Manager* are not subject to the clause 61.3 time bar.

101. Not accepting the design (clause 60.1(9))

QUESTION

I am trying to understand clause 60.1(9) and think of an instance when this could apply – can you advise?

ANSWER

Let us say that we have asked a *Contractor* to build a new football stadium. In the Scope, we have asked it to design all the *works* (clause 21.1 refers). We then require it to submit the steelwork design (clause 21.2). In the Scope, we do not impose any constraints or mention the potential for future expansion to the football stadium. When the steelwork drawing comes through, we then realise, based on the structure, that it would prevent future expansion. Our only two reasons for not accepting the design under the ECC (mentioned under clause 21.2) are that it does not comply with the Scope or the applicable law – we therefore cannot 'not accept' based on those two reasons in this instance, so we must 'not accept' for 'a reason not stated in this contract'. We can do so, but then, as is fair, this would give rise to a compensation event.

102. Tests and inspections that the Scope requires (clause 60.1(10))

QUESTION

I am a frustrated *Supervisor* on an Option C ECC project! The *Contractor* is hardly doing any tests and inspections – only the bare minimum required by law, and I suspect that there may be Defects. What can I do? Can I request a search for suspected Defects?

ANSWER

This all comes back to drafting the Scope properly at the outset – it sounds like you were not given the opportunity to input into this. Clause 41.1 states that the test and inspection clauses only apply to those required by the Scope or applicable law.

You can get the *Project Manager* to instruct a change to the Scope (clause 14.3), but this will be a compensation event (clause 60.1(1)). An alternative approach would be to get the *Project Manager* to notify an early warning and get involved in discussions about the potential impact of such a change.

If you do suspect Defects, then you have the power as the *Supervisor* to instruct a search. However, if none is found, it is a compensation event.

The best approach here is to deal with things proactively and perhaps change the Scope to reflect the tests and inspections that you feel are appropriate (this needs to be by means of the *Project Manager* giving an instruction). Clause 41.3 then requires the *Contractor* to inform you of these before they are undertaken and the result afterwards.

103. Unnecessary time for tests and inspections (clause 60.1(11))

QUESTION

We are a *Contractor* on a sports centre project, and the *Supervisor* is taking what we believe to be an unnecessary amount of time to do its tests and inspections. Where do we stand with this?

ANSWER

Ideally, the Scope would have stated the tests and inspections that are required (clause 41.1) and how long they will take. The tests and inspections normally fall into three categories: those that the *Contractor* is required to undertake, those undertaken by the *Supervisor* and those that the *Supervisor* wishes to witness. If a timescale is not stated in the Scope, then it may come down to interpretation and common practice. If you feel that this is a compensation event, then you should notify as such (clause 61.3 requires this to be undertaken within 8 weeks of becoming aware). As it seems to be a repeat problem/trend, we would also notify an early warning and try to resolve proactively.

104. Interpreting physical conditions (clause 60.1(12))

QUESTION

We have a *Contractor* on Site under ECC Option A. Access to the *works* involves driving along a quay, down a ramp and onto the beach – the project involves undertaking jetty repairs to one of our docks. The *Contractor* has recently notified a compensation event under clause 60.1(12) for a 'max. headroom' goal post steel frame, which prevents certain-sized Equipment using the ramp. We did not mention this in the Site Information, but the *Contractor* was allowed to undertake a Site visit as part of the tender. Can you advise where we stand with this?

ANSWER

The two clauses that need to be in conjunction with one another are clauses 60.1(12) and 60.2. Clause 60.2 explains that 'in judging the physical conditions' the *Contractor* is assumed to have taken into account information obtainable from a visual inspection. So, just what is contained in the Site Information is not the whole picture. If something could have been seen from a visual inspection (and we are sure that the steel frame could), then it would not be a compensation event.

105. Underground mine workings (clause 60.1(12))

QUESTION

We are a local authority about to use an Option A ECC contract to procure the installation of a flood wall in a coastal area. The area is notorious for underground mine workings, but we do not have accurate records as they were not kept in the past. We do not want, on discovery of underground mine workings, this to be a source of debate nor do we want the tenderers to price for this and for it still to be debated as a compensation event. Do you have any suggestions? We want to be a fair procurer and not force undue risk onto the tenderers.

ANSWER

Two options exist: you could either introduce an Option Z clause to clarify the definition or add an additional compensation event in Contract Data part one. The additional compensation event is perhaps best in this situation. This could be stated as 'Discovery of underground mine workings' (or perhaps clearer wording). If this occurs, it would then be a compensation event under clause 60.1(21). This will ensure that the *Contractors* do not price this liability in the tender – the only disadvantage is that if the liability occurs it will be a compensation event. The main principle is that it is clear and, as per your objective, fair.

106. What are physical conditions? (clause 60.1(12))

QUESTION

In an ECC contract, what is the accepted definition of 'physical conditions' in clause 60.1(12)? Are there circumstances in which physical conditions under the ECC could cover uncharted services?

ANSWER

A glance at a dictionary says that 'physical' in the sense used here means 'of material things or nature'. Physical conditions would include artificial obstructions. A pipe or cable, whether charted or uncharted, is a physical condition, and we have never heard anybody seriously suggest otherwise.

107. Weather risk (clause 60.1(13))

QUESTION

We are a large contractor and want to understand the risk as defined under clause 60.1(13) of the ECC. Can you clarify?

ANSWER

Scientific assessment of weather is included in the ECC. Firstly, the *weather measurements* to be recorded are those detailed in Contract Data part one, so these need to be reviewed for each tender (note that wind is not included by default). Only the *weather measurements* detailed in Contract Data part one are relevant in assessing a compensation event. Each month is considered in isolation, so bad weather spanning 2 months would be reviewed as two separate compensation events. The *weather measurements* are also only relevant if taken before the Completion Date for the whole of the *works* – so if you record data late, then all weather is the *Contractor's* risk.

Depending on who is providing the weather data (which will be defined in Contract Data part one), it may be calculated in different ways. If it is the UK Met Office, then a long-term 30-year period is typically assessed, and the worst three measurements averaged to give the weather shown to occur 'on average less frequently than once in ten years'. This is also known as the 1 in 10 return value – details of ECC weather data and how they are calculated are on the Met Office website. The good news is that weather can be scientifically assessed both at tender and for compensation events. The background risk, however, is not insignificant, and so needs to be considered carefully in your tender programme and price.

108. Take over (clause 60.1(15))

QUESTION

We are delivering a data centre and need to get into some of the rooms to store kit that has arrived early. This is under ECC Option C. Are we allowed to start using parts of the building before the *Contractor* has finished?

ANSWER

Yes, clause 35.2 allows the *Client* to use part of the *works* before the Completion Date – so you have the authority. However, this will be a compensation event (clause 60.1(15)). As soon as the *Client* begins using the *works*, then take over is said to have occurred – this is then certified by the *Project Manager* within 1 week.

Also, from take over, loss or damage then becomes the *Client's* liability (clause 80.1). You would be well advised to discuss this with the *Contractor* as early as possible.

109. Assumptions made by the *Project Manager* (clause 60.1(17))

QUESTION

We have some complex compensation events on our ECC Option C project. As the *Project Manager*, I do not want the *Contractor's* assessments to be full of risk – how can I deal with this?

ANSWER

At a macro level, *Client's* risks (Contract Data part one) change the risk profile of the contract. On a compensation event by compensation event basis we can add assumptions as a *Project Manager* under clause 61.6. The original compensation event is not then revised if an assumption is wrong (clause 66.3). This becomes a separate negative or positive compensation event to correct the assumptions (clause 60.1(17)).

This an excellent process of getting better value from compensation events, fair risk allocation and enabling complex compensation events to be assessed.

110. Prevention (clause 60.1(19))

QUESTION

Does Subcontractor insolvency come under clause 60.1(19) of the ECC?

ANSWER

Clauses 19.1 and 60.1(19) need to be carefully considered. They define an event that stopped the *Contractor* completing the *works* or completing the *works* by the date shown on the Accepted Programme, and that neither Party could predict or prevent. The wording also states that it would have been unreasonable for the *Contractor* to have allowed for the event.

Subcontractor insolvency may not fall into this category – it happens relatively frequently, unfortunately. This clause is designed not to pre-empt a defined list of issues but rather describe the scenario. For a compensation event to be successful under clause 60.1(19), it would need to pass all of the tests contained within the clause.

111. Poor Site Information (clause 60.2)

QUESTION

What happens if the *Client's* Site Information is ambiguous or inconsistent?

ANSWER

Site Information is produced by the *Client*. If it is found to be ambiguous (vague) or inconsistent (contradictory), then the *Contractor* is favoured (clause 60.3) when assessing a compensation event under clause 60.1(12) for physical conditions.

112. Mistakes in a Bill of Quantities (clause 60.6)

QUESTION

We have an ECC project under Option B. There is a requirement in the Scope for aerial video surveys undertaken by drones. There is no Bill of Quantities item for video surveys. Are we entitled to a compensation event by way of the following process?

1. Inform the *Project Manager* of an ambiguity or inconsistency in or between the documents in accordance with clause 17.1.
2. The *Project Manager* must correct mistakes in the Bill of Quantities under clause 60.6, and such corrections are compensation events.
3. Clause 60.7 says that when assessing such a compensation event, the *Contractor* is assumed to have taken the Bill of Quantities as correct.
4. Therefore, we are entitled to a compensation event for the cost of the video surveys plus the Fee.

Note: all of this has just arisen, and there has been no correspondence to date with the *Project Manager* on the issue.

ANSWER

Clause 60.6 only applies to certain 'mistakes' – that is, when they are departures from the rules and divisions set out in the *method of measurement*. The answer to your question therefore depends entirely on what the *method of measurement* says about such videos. If they are included, either explicitly or implicitly, as part of the item coverage, then there is no mistake. If they are not covered or are mentioned as being required to be measured separately, then there is a mistake.

Clause 60.7 only applies to how the compensation event is assessed, not whether it is a compensation event in the first place. So, if it is a compensation event under clause 60.6, then, under clause 60.7, it is assessed assuming that you did not allow for it in your Prices.

113. Error in the Bill of Quantities caused by the *Contractor* (clause 60.6)

QUESTION

In a project using ECC Option B, a *bill of quantities* was issued to tenderers to price. The successful *Contractor* returned the priced *bill of quantities* (which of course became the Bill of Quantities) but in a different format so that it aligned with its subcontract packages. In rearranging the tender *bill of quantities* items, the *Contractor* missed off some measurable *works*, according to the *method of measurement*. The Scope has not changed, but is it correct to assume that this becomes a compensation event under clause 60.6, even though it is an error originally made by the *Contractor*?

ANSWER

You are correct – this would be a compensation event under clause 60.6. Furthermore, when assessing this compensation event, the *Contractor* is assumed to have taken the Bill of Quantities as correct (clause 60.7). In the circumstances, the changes made by the *Contractor* to the Bill of Quantities were, in effect, a qualification to its tender. Your client should have treated it as such and checked it very carefully before accepting the tender. Once accepted, the risks of such errors pass to the *Client*.

114. The difference between clauses 61.1 and 65.1

QUESTION

What is the difference between a compensation event under clauses 61.1 and 65.1?

ANSWER

Clause 61.1 deals with the requirement of the *Project Manager* to notify certain compensation events in certain instances. If this has arisen due to, for example, a clause 14.3 *Project Manager's* instruction to change the Scope, then the *Contractor* must obey this instruction under clause 27.3 and put this change into effect. This is a clear 'green light' to undertake whatever has been requested. An instruction to provide a quotation will be included in the notification of this compensation event (clause 61.2), and this will then follow the compensation event assessment process.

Clause 65.1 is not a compensation event – it is an instruction to submit a quotation for a proposed instruction. The proposed change is thus not put into effect – we are effectively requesting a quotation, and then we shall decide in due course.

An obvious comment, but these notifications need to be clearly communicated (citing the correct clauses) in order to avoid confusion.

115. The time bar under clause 61.3

QUESTION

As a *Contractor* under an ECC Option B contract, we have notified the *Project Manager* of poor ground conditions under clause 15.1. Should it notify a compensation event, or do we?

ANSWER

Early warnings and compensation events are separate procedures. The easiest way of thinking about these is that there is no time or money in early warnings – only risk reduction. If, as a *Contractor*, you wish to be compensated with time and/or money, then the only way to do this is through a compensation event. Clause 61.3 clarifies that if you do not notify within 8 weeks of becoming aware, you will lose entitlement. This is referred to as the time bar. The exception to the time bar is if the event arose from the *Project Manager* or *Supervisor* giving an instruction, issuing a certificate, changing an earlier decision or correcting an assumption.

The clue to which of the 21 core compensation events may be time barred is the wording. For example, clause 60.1(1) would not be time barred as it would arise from an instruction from the *Project Manager*. However, clauses 60.1(2) and 60.1(12) would be, and so on.

116. *Project Manager's* failure to respond (clause 61.4)

QUESTION

We have a situation in which the *Project Manager* is not responding to our notifications of compensation events under Option A of the ECC. As a *Contractor*, we need to get the financial and programme impact of compensation events agreed. What can we do to progress these?

ANSWER

Clause 61.4 provides 1 week for the *Project Manager* to respond to a notified compensation event (or longer time if agreed with the *Contractor*). Ultimately, if the *Project Manager* does not respond, the *Contractor* may (which means it is discretionary) notify the *Project Manager* of its failure. If the *Project Manager* fails to respond within a further 2 weeks, its failure is treated as an instruction to provide a quotation.

It is always best to agree collaboratively if possible – perhaps consider notifying an early warning and bring the backlog to a head?

117. Link between early warnings and compensation events (clause 61.5)

QUESTION

Our *Contractor* failed to notify an early warning for a matter that was ultimately a compensation event. While undertaking refurbishment work in our office block, it discovered a damp patch in several ceiling tiles. Without an early warning, it took it upon itself to replace the whole ceiling. However, we are frustrated as the ceiling was going to be taken out by another contractor in our employment and a painted finish applied to the services – effectively doing away with the suspended ceiling. Where do we stand with this – do we have to pay?

ANSWER

There are several clauses that are linked here (clauses 15.1, 61.5 and 63.7). Clause 15.1 obliges the *Contractor* to notify an early warning. Clause 61.5 states that, on notification of a compensation event, if you believe that an early warning could have been given, then you notify this decision. This is then linked to clause 63.7, which states that, if notified under clause 61.5 (so this becomes a condition precedent), it is assessed as if the early warning had been given. So, in your case you would assess the compensation event as nothing, as your other contractor would have dealt with it. You have not stated which main Option you are using, but if you are using Option C, D, E or F, then the excess cost will also be disallowed (Options C–E clause 11.2(26) and Option F clause 11.2(27)).

118. Compensation events after Completion (clause 61.7)

QUESTION

Our *Contractor* has successfully completed a new school for us. We certified Completion 3 months ago, and the *defects date* is 52 weeks after Completion. Are we able to get it back to do additional work as a compensation event?

ANSWER

In short, yes. Clause 61.7 allows for compensation events to be notified up to the *defects date*. Any instruction given under clause 14.3 must be obeyed under clause 27.3. The only downside is that the cost will be more expensive than when the *Contractor* was originally on Site. The clause 63.1 forecast of Defined Cost will include any mobilisation costs, for example.

119. Additional work after Completion (clause 61.7)

QUESTION

I am a *Project Manager* on an Option B ECC project. The project achieved Completion 3 months ago for which I issued the Completion certificate. The *Client* now wants some additional work undertaken but I cannot find a clause that allows me to instruct this.

ANSWER

The ECC provides provision for this by the usual clause 14.3 instruction to change the Scope. Under clause 61.7, a compensation event can be notified at any point up until the *defects date*. The only disadvantage of notifying compensation events after Completion is that the *defects date* will not be extended, and the compensation events will be more costly – clause 63.1 explains that they will be based on a forecast cost. Inevitably, the cost to the *Contractor* will typically be greater once it has left the Site.

120. Obtaining different quotes for the same compensation event (clause 62.1)

QUESTION

We have a compensation event on an Option D ECC project. Are we able to get different quotations for different methods of dealing with a particular problem?

ANSWER

Yes, clause 62.1 enables the *Project Manager* to obtain different quotations for the same compensation event. The wording used is 'After discussing with the *Contractor* different ways of dealing with the compensation event', so it requires the *Contractor* and *Project Manager* to work together closely and develop sensible, practical approaches.

121. Updating for remaining work (clause 62.2)

QUESTION

I am a bit concerned on our ECC Option A project. Clause 63.5 states that we use the latest Accepted Programme and then assess the impact of the compensation event on planned Completion. However, our problem is that the *Project Manager* has not accepted the programme for 6 months. Surely this cannot be right, as we are looking at the impact on an out-of-date programme?

ANSWER

The ECC clauses cannot be read in isolation, and this instance is a good example. Clause 62.2 needs to be read in conjunction with clause 63.5. Clause 63.5 requires the latest Accepted

Programme to be used, for this to be updated for 'remaining work' (clause 62.2) and then the impact assessed on planned Completion (back to clause 63.5). Both clauses are used by the *Contractor* in its quotation for a compensation event. The same methodology is used if it is the *Project Manager* that assesses – clause 64.2 requires the programme to be updated for 'remaining work'.

122. Proposed changes to which programme? (clause 62.2)

QUESTION

We are the *Client* and are using ECC Option C. A *Project Manager's* instruction has been issued changing the Scope. The *Project Manager* has also notified the *Contractor* of the compensation event and instructed the *Contractor* to submit quotations at the same time, as required in clauses 61.1 and 61.2. The *Contractor* submitted a quotation for the compensation event within 3 weeks. The *Project Manager* has instructed the *Contractor* to submit a revised quotation (clause 62.4), which the *Contractor* has also submitted within 3 weeks. The *Contractor* has included within its quotations proposed changes to the Prices and a delay to the Completion Date (as planned Completion is delayed). In accordance with clause 62.2, the *Contractor* has included alterations to the Accepted Programme in its quotation.

Should the *Contractor* base the proposed change to the Completion Date on (*a*) the Accepted Programme at the time of issue of the *Project Manager's* instruction and request to submit quotations or (*b*) the Accepted Programme at the date of the *Contractor's* alternative quotation submission?

ANSWER

This is one of the problems with delaying the process; things have now moved on. The *Contractor* should have used the Accepted Programme as *(a)*, because that was what was current at the dividing date (clauses 63.1 and 63.5).

123. Accepting quotations in a timely manner (clause 62.3)

QUESTION

We are the *Contractor* under an ECC Option C contract. We have submitted several quotations to the *Project Manager*, and we have agreed to relax the response period for the *Project Manager* to accept our quotations. We now find ourselves in a position in which we are agreeing quotations with our Subcontractors before we have received acceptance of our quotations. There have been several occasions where the *Project Manager* assessed compensation events at a lesser value than we had already agreed with the Subcontractor. What should we do? Also, if a Subcontractor's actual Defined Cost is greater than how the *Project Manager* assesses our compensation event, is the difference a Disallowed Cost?

ANSWER

The first point is to question why you made such an agreement with your *Project Manager*. If it causes such problems, then insist the *Project Manager* deals with your quotations within the timescales set out in clause 62.3. If the *Project Manager* fails to do so, then notify the *Project Manager* of that failure under clause 62.6.

If the *Project Manager* continues to fail to reply for 2 weeks after that notification, then your original quotation will be treated as accepted (clause 62.6) and therefore implemented (clause 66.1), after which only the *Adjudicator* can change it. The timescales in the ECC are deliberately written so that you should not need to get yourself into this position.

The answer to the second question is that this excess amount is not a Disallowed Cost, simply because it is not stated within the list of Disallowed Cost items listed in clause 11.2(26).

124. Failing to respond (clause 62.3)

QUESTION

Our *Project Manager* on an Option A ECC highway project constantly fails to respond to compensation events that we have quoted. What should it do on receipt of a quoted compensation event, and what are the remedies in the contract if it does not respond?

ANSWER

On receipt of the *Contractor's* quotation, the *Project Manager* has three options only (clause 62.3). It is required to accept the quotation, request a revised quotation (stating reasons as per clause 62.4) or decide to make its own assessment. It needs to decide which of these three actions it will take within 2 weeks.

Failure of the *Project Manager* to respond to the *Contractor* may (discretionary) result in the *Contractor* notifying it of its failure (clause 62.6). Failure of the *Project Manager* to respond within 2 weeks of this is then treated as acceptance of the quotation. This is not revised (clause 66.3). The ultimate remedy in the contract is that continued failure of the *Project Manager* to respond will result in the compensation event being implemented.

The *Project Manager* and *Contractor* should really be working closely in order to gain agreement before submission and resolve queries. Regular meetings are a practical way of keeping on top of the process.

125. What programme should we use when assessing the impact of compensation events? (clause 63.1)

QUESTION

In assessing the impact of a compensation event on planned Completion, the ECC states the latest programme is to be used. How do we work out what is the date of the latest programme? Is it the date when the *Contractor* became aware of the event itself, or the date when the *Contractor* notified the compensation event?

ANSWER

The answer to your question lies in the last part of clause 63.1 read together with clause 63.5. If it was a compensation event that the *Project Manager* should have notified (see clause 61.1 for these events), then it is the Accepted Programme current at the dividing date – that is, the date of the communication (i.e. the instruction) by the *Project Manager*. For all other compensation events, the dividing date is the date of the notification of the compensation event by the *Contractor*, and you use the Accepted Programme current at that date. You need to bear in mind with the latter that compensation events are subject to an 8-week time bar in clause 61.3.

126. Using the Short Schedule of Cost Components (SSCC) for Subcontractor's costs (clause 63.1)

QUESTION

Our queries relate to a project using ECC Option A. When assessing compensation events involving Subcontractors, we understand you use their costs including their mark up for their overheads and profit.

Our questions are the following.

1. Do you have any thoughts on this?
2. What process should be followed if the *Contractor* has not used the NEC subcontracts?

ANSWER

For Question 1, you need to consider the (forecast) payments to Subcontractors for work that is subcontracted (item 41 of the SSCC). Of course, this needs to be an open market or competitively tendered price in accordance with clause 52.1, which applies for all such Defined Cost where these are not stated in the Contract Data.

Regarding Question 2, you follow exactly the same process, in that you have to forecast what the effect will be on Defined Cost. If the *Contractor* has chosen not to use an NEC subcontract, that means it may be paid on an entirely different basis to that which it pays its Subcontractor – which may or may not be beneficial to the *Contractor*. That is its concern, not yours. You (and the *Contractor*) must forecast the cost by using the rules in the SSCC.

127. Forecasting the Defined Cost (clause 63.1)

QUESTION

How should compensation events be valued? Is it based on the actual cost incurred by the *Contractor* plus Fee? What happens if the estimate is wrong – is it revised based on the actual cost?

ANSWER

This aspect of the contract can cause confusion in practice.

Clause 63.1 states that the compensation event assessment is based on a mixture of actual Defined Cost of the work done by the dividing date, a forecast of the work not done by the dividing date plus the resulting Fee.

Note that Defined Cost is a defined term and therefore means different things in different main Options. The 'switch' between actual and forecast is not when the person assessing the event gets around to doing it. It is based on the statements in clause 63.1.

If the compensation event results from the *Project Manager* or *Supervisor* giving an instruction or notification, issuing a certificate or changing an earlier decision, the dividing date between actual and forecast is the date of that communication (note that the compensation event may be notified after this point). That is, in all other cases the date of the notification of a compensation event becomes the switch from actual to forecast – so as soon as the *Contractor* notifies a compensation event, this would require the assessment to be a forecast from that point.

Compensation events under the ECC are therefore based on largely a forecast/estimate rather than on actual cost. There are other clauses that need to be taken into consideration, such as clause 63.8, which allows the *Contractor* to include time and cost risk allowance. It is all about having a reasoned/fair estimate.

128. Using the *fee percentage* (clause 63.1)

QUESTION

We are working with the *Contractor* to agree costs for compensation events and early warnings under ECC Option A. Under clause 61.3, the *Contractor* has notified a compensation event after acceptance of a revised programme by the *Project Manager*. The *Contractor's* quotation for the compensation event includes Defined Cost for the extended programme, which includes Site accommodation and running costs, key personnel (contracts manager, quantity surveyor, Site engineer etc.), Site hoarding and so on. After calculating this, the quotation includes a *fee percentage* uplift. Can this uplift be used, and would not this cost normally be included in the Fee anyway, as per clause 52.1?

ANSWER

We are not sure why you consider that all these costs should have been included in the Fee because clause 52.1 does not support that view. Defined Cost is a defined term – for Option A see clause 11.2(23). That, in turn, refers you to the SSCC, which sets out what will be paid for as part of Defined Cost. A compensation event is assessed by its Defined Cost and then the Fee is added (clause 63.1). If it can therefore be shown that one of the costs in the SSCC has been increased because of the compensation event, then it will be assessed in the way set out in the SSCC. Conversely, only if a cost is not listed in the SSCC will it not be part of Defined Cost, and therefore be part of the Fee under clause 52.1.

So, you need to look at the SSCC in detail to decide which of these costs are paid. For example, the cost of people is included in the People Rates in SSCC item 11, if this meets one of the three criteria in item 1. If, for example, the quantity surveyor spends 2 days a week on Site and the other 3 days working at head office, then 2 days of that person's People Rates is allowed, assuming that person's normal place of working is elsewhere. Equally, from SSCC item 2, the cost of the *Contractor's* Equipment, including accommodation, is included along with Site hoarding, which is also Equipment. The *Contractor* is correct to provide its quotation based on a forecast of what these costs will be (clause 63.1). In NEC contracts, you do not wait until the costs have been spent – you forecast what they should be, assuming that the *Contractor* acts promptly and competently (clause 63.9) and allowing for risks that have a significant chance of occurring (clause 63.8).

What does worry us here, though, is that you say that the *Contractor* has notified a compensation event because the *Project Manager* has accepted a revised programme. That cannot be the case. The list of compensation events is set out in clause 60.1, and there is not one that says 'the *Project Manager* has accepted a programme that shows late Completion'. On the other hand, if a listed compensation event means planned Completion is now later than planned Completion shown on the Accepted Programme, then the Completion Date will be extended (clause 63.5). That will, however, be because of the underlying compensation event, not because the *Project Manager* has accepted the subsequent programme.

129. Recovering the cost of additional Site staff (clause 63.1)

QUESTION

As the *Contractor* with a compensation event arising in an ECC Option B contract, are we able to claim for additional site staff costs (i.e. contracts manager and quantity surveyor) to cover the pricing and delivery of a compensation event, even when the overall duration has not increased?

ANSWER

Regarding the pricing of the compensation event, this might be recoverable if you can demonstrate that the compensation event has an effect on the Prices (clause 63.1). Compensation events are assessed based on their effect on Defined Cost (clause 63.1) but remember that Defined Cost is a defined term (see clause 11.2(23) and the SSCC).

In the SSCC, you will see you only get paid for people when they comply with one of the three bullets in item 1, and only if your cost of them has increased. Just because your site agent or perhaps engineer worked on them does not mean that your cost increased, if they were going to be there anyway. They just maybe worked a bit harder. On the other hand, if the time you are on Site was extended because of the compensation event, then their Defined Cost would increase. Or if the compensation event was such as requiring an additional engineer to run it, again the cost of that would be included. Just because somebody works on it does not mean your cost increases.

130. Charging for staff time in a compensation event (clause 63.1)

QUESTION

I have the following questions about an ECC Option A contract.

1. Can a *Contractor* charge for its quantity surveyor's time in preparing a compensation event quotation?
2. Can a *Contractor* charge for its quantity surveyor's time in sorting out quotes and placing orders necessary for a compensation event quotation?
3. Can a *Contractor* charge for its Site engineer's time in working on a compensation event in its quotation for a compensation event?

ANSWER

The answer to your first question is 'possibly', if you can demonstrate that there is an effect of the compensation event on the Defined Cost (clause 63.1).

The answer to your other two questions is a bit more complicated and will depend on the circumstances. The value of the compensation event is based on its forecast effect on Defined Cost (clause 63.1). So, the *Contractor* must show that Defined Cost for these people increased. If they worked on them and were going to work on Site anyway, then the *Contractor* cannot show that Defined Cost for them has increased. All that has happened perhaps is that the people have worked a bit harder. On the other hand, if the compensation event delays Completion by, say, a week, then the *Contractor* and these people will be on Site for another week and their Defined Cost will increase, and that increase is included in the valuation of the compensation event. Equally, if the compensation event is of such a size that it is forecast that an additional engineer will need to be brought in to manage it, then again that will increase Defined Cost, and that will be included in the valuation of the compensation event. So, the answer to the second and third questions is 'it depends'.

131. People costs in a compensation event quotation (clause 63.1)

QUESTION

As part of an ECC Option A contract, when assessing compensation events, the *Contractor* is including time for its quantity surveyor, who is based at its head office (and so outside the Working Areas), and its senior engineer, who is based full time at the Site. The *Contractor* is tending to charge a nominal amount of two hours for the quantity surveyor and four hours for the senior engineer for every compensation event. Can you confirm if the *Project Manager's* following interpretations of the contract are correct?

1. Under SSCC item 1 (people), the first category of people excludes *Contractor's* staff who are working at the head office. Therefore, the *Contractor* cannot claim time for the quantity surveyor, as their time is covered by the Fee.
2. SSCC item 1 includes people whose normal place of working is within the Working Areas, so the senior engineer is a role that can be charged. However, in this instance the senior engineer is already assigned full time to the contract, so the *Client* is already paying for 100% of their time. Therefore, if the senior engineer is directly working on an activity that is the subject of a compensation event, the *Contractor* can claim for their time.
3. If the senior engineer is not directly working on an activity that is the subject of a compensation event, but the compensation event delays planned Completion, the *Contractor* can claim the cost of its time for the delay period only.
4. If the senior engineer is not directly working on an activity that is the subject of a compensation event and the compensation event does not delay planned Completion, the *Contractor* cannot claim the cost of their time.

ANSWER

The assessment of a compensation event is based on what effect it is forecast to have on Defined Cost. The answer is not quite as simple as you have set out. The first thing to say is that the cost has to increase. If the *Contractor* has an engineer or quantity surveyor on Site anyway and it uses that person to manage the compensation event, then Defined Cost has not increased, so it is not included. Of course, if the compensation event is such as to lead it to need another engineer on Site (either because it is large, or there have been a lot of other changes and this one 'breaks the camel's back'), then that will be included. Or if it delays its work so that the engineer is on Site for longer than they would ordinarily have been, then that will be allowed for. Regarding people whose normal place of working is not within the Working Areas and who are working outside the Working Areas, they are not included as part of the definition of Defined Cost, and, under clause 52.1, they are assumed to be included in the Fee. So, your interpretation 1 is correct; 2 is only correct to the extent that its cost can be shown to be forecast to increase; and 3 is correct, as is interpretation 4. The only exception is where the people are listed in the Contract Data in accordance with the third bullet of SSCC paragraph 1.

132. Are compensation event quotations based on actual cost? (clause 63.1)

QUESTION

Our queries relate to a project using ECC Option A.

1. Clause 63.1 states: 'For a compensation event that arises from the *Project Manager* or the *Supervisor* giving an instruction or notification, issuing a certificate or changing an earlier decision, the dividing date is the date of that communication'. If the *Contractor* does not submit a quotation until well after the *works* are complete, should the quotation be based on the actual cost or on forecasts of the actual cost of the *works*?
2. If the *Project Manager* does not instruct the *Contractor* to submit quotations until after additional *works* were carried out, due to them not being brought to its attention by the *Contractor* until after they were complete, should the quotation be based on the actual cost or on forecasts of the actual cost of the *works*?

ANSWER

The answer to Question 1 is that you assess using a forecast, no matter when that assessment is being done. And that forecast assumes the *Contractor* acts promptly and competently (clause 63.9) and allows for risks that have a significant chance of occurring (clause 63.8).

Regarding Question 2, the answer will depend on the reason for the compensation event. If it arises from a *Project Manager's* instruction to change the Scope, then the obligation is on you to notify the compensation event and instruct a quotation (clauses 61.1 and 61.2). So, you will use a forecast because the dividing date is the date of that communication (the instruction, clause 63.1). If it is one that the *Contractor* should notify, then you first have to consider whether the time bar in clause 61.3 applies. If it does not, then the dividing date is the date of notification of the compensation event, and you would probably use actual Defined Cost in this instance (clause 63.1), but again by assuming the *Contractor* acted promptly and competently.

133. Compensation event quotations and Subcontractors (clause 63.1)

QUESTION

We (the *Client*) are using ECC Option C and have a quotation for a large compensation event for additional steelwork, which we are worried is excessive. We are looking for confirmation on using a transparent method to review the proposed costs and accept them (or not) in accordance with the contract. The quotation includes both costs for manufacture and large amounts for Site labour, which we cannot see any justification for. We understand the vast majority (if not all costs) stem from the steelwork fabrication Subcontractor, which will both supply the steelwork and fit it on Site. The quotation includes large sums under people costs. Looking at the Schedule of Cost Components (SCC) descriptions and guidance, we consider that this should only be the *Contractor's* directly employed staff, not subcontracted people. We do note that the SSCC people costs do, however, allow for Subcontractor people costs on Site.

Should the *Contractor* be using the SCC for quotations involving Subcontractors' costs or should we be agreeing to use the SSCC to break down all compensation events involving Subcontractors? Or should it simply provide the Subcontractors' quotations to accompany its own quotation and add on its *fee percentage* for Subcontractors? If the latter, how can we have sight of where the costs are going so that we can consider whether the costs are justified? Can we ask that the steelwork fabricator's quotation be broken down – and if so, to what level of detail?

ANSWER

This compensation event is valued using Defined Cost plus Fee (clause 63.1). The definition of Defined Cost in ECC Option C is set out in clause 11.2(24), which directs you to the SCC. Here, you will see that subcontracted work is included at paragraph 4.

If one Subcontractor will supply and erect the steelwork, then all its costs will be recovered through paragraph 4 (Subcontractors) of the SCC. However, if the steelwork is supplied by one firm, separately to the erection, then these costs would be captured under paragraph 3 (Plant and Materials) of the SCC. The erection costs would then be captured either under paragraph 1 (People) or paragraph 4 (Subcontractors).

However, the costs are categorised; remember the introductory paragraph to the SCC. Any cost is only included in one cost component, there is no double counting.

As to the amount the *Contractor* pays its Subcontractor, that will depend on the terms of the subcontract it is under, which you should have seen when accepting that Subcontractor (clause 26). If, for example, the subcontract values variations based on rates, then those rates are used, and both the *Contractor* and you are stuck with them and cannot ask for any further substantiation. On the other hand, if the contract is based on assessing Defined Cost, then you are both stuck with that.

In all of this it is necessary for the *Contractor* to provide enough information to satisfy you that that is what it is going to pay. Remember that this is an 'open book' contract, so you are entitled to see whatever the *Contractor* has on that (clauses 52.2 and 52.3). You are also entitled to see, in the case of a subcontract value based on Defined Cost, that the assessment has been carried out at open market or competitively tendered rates (clause 52.1). This may well require a detailed breakdown. It is not possible to tell you how much information you will need in order to do that, as it will depend on the circumstances. This is something that you need to discuss with your *Contractor*.

134. People not working in the Working Areas (clause 63.1)

QUESTION

The Working Areas in our ECC contract are 'the Site'. We are applying for compensation events as per the SSCC, which includes project managers, quantity surveyors, buyers and so on. The *Project Manager* is stating that as these individuals are working at the head office and not on Site, they are included within the Fee and cannot be separately charged for.

Our queries are, first, as these individuals are listed in the SSCC through the People Rates, can they be charged for even if they are not working in the Working Areas? Second, it is my understanding that the process with compensation events is to put you in a position that you would have been in had the compensation event never occurred in the first place. Therefore, I believe I should be entitled to charge for individuals who assisted in preparing quotations for and undertaking compensation events.

ANSWER

The *Project Manager* is partly correct for two reasons. The first is the one it has given you: some people are only paid for when and if they are working in the Working Areas (see the bullets in item 1 of the SSCC). Any other costs are assumed to be in your Fee (clause 52.1). People whose normal place of working is within the Working Areas are paid for the work they do on the contract, wherever that takes place. People whose normal place of working is outside the Working Areas only get paid for the time they spend within the Working Areas.

There is a third category of person, those people who are named in the Contract Data as being paid for when outside the Working Areas. This list of people is separate from the People Rates that you mention. So, the *Project Manager* is only partially correct on this one. It depends on the category of person that you are considering.

In answer to your second question, the contract is intended to be neutral on this, but it assumes that you have allowed enough in the Fee you quote to cover what you are not going to get directly paid for. If you have underpriced the Fee, then this will be at your risk. A further reason is that the value of compensation events will be based on what your Defined Cost has increased by (clause 63.1). Just because an employee works on something does not necessarily mean that has increased your Defined Cost if you would have employed that person anyway.

135. Extensions of time (clause 63.5)

QUESTION

How are extensions of time claims dealt with in the ECC? Do you compile these at the end of the project when you know the overall impact of all the variations?

ANSWER

In short, no. Each compensation event needs to be assessed within the stated timescales, and both the time and cost impact are assessed jointly. The *Contractor* has 3 weeks to assess both (clause 62.3) or a later period agreed with the *Project Manager* (clause 62.5).

The latest Accepted Programme is adopted (clause 63.5), then updated for remaining work (clause 62.2). Next, whether planned Completion (i.e. when the *Contractor* plans to complete) is affected is assessed. If it is affected, then by whatever duration is added to the Completion Date. The ECC provides very clear rules for assessing the programme impact within the core clauses. There is no room for assessing at the end of the project – clause 66.3 clearly states that assessments are not revised, and this is applied to both time and cost, so there is a 3-week window to assess and ensure that this is accurate.

136. Dominant cause of delay (clause 63)

QUESTION

A design-and-build project is being carried out under ECC Option A. The project is linear, involving access onto adjoining greenfield lands owned by different people to construct a roadway. Although there are *sectional* Completion Dates, there is only one *access date* included in the contract. The clause 60.1(13) compensation event for weather has been deleted.

The project has suffered an overall delay of 10 weeks. Firstly, 3 weeks of this delay was experienced due to inclement weather. Shortly afterwards, the *Contractor* progressed the *works* such that access was required from one piece of land onto another. The *Client* advised the *Contractor* to demobilise, as agreements had not been fully concluded with a particular landowner, and centred around more land-take now required than had previously been planned. Due to this, a further delay of 7 weeks followed where the *Contractor* was in delay due to the stand-down.

The *Contractor* maintains that had there been no weather delay the non-access delay would have been realised earlier, yet still not resolved until the time it was. The basis for this point of view is the fact that the *Contractor's* design drawings, and revisions thereto at the request of the *Project Manager*, were submitted in good time (and accepted) well ahead of the planned start of *works* onto the affected lands; therefore, the *Client* had every opportunity of agreeing any additional land-take required with the landowner in good time to obviate the subsequent delay and ahead of the weather delays. The *Contractor* maintains that the dominant delay is the non-access. Currently, the *Project Manager* recognises the 7-week delay only as a direct effect of the non-access but is holding the *Contractor* liable for the 3-week weather delay and delay damages.

Is the *Project Manager* correct in this approach, and what is the protocol for establishing the dominant delay?

ANSWER

Delays that the *Contractor* knows about must be dealt with in each programme that the *Contractor* issues for acceptance. We assume here that they are to be submitted monthly. The *Contractor* therefore should have included in its next programme issued for acceptance the effect that these 3 weeks have had on the actual progress and the order and timing of the remaining work. That may show planned Completion being moved back by 3 weeks or it may show the remaining *works* being reprogrammed to catch up the 3 weeks. Once this programme is accepted, it becomes the Accepted Programme (clause 11.2(1)) and is used to assess future compensation events, including this one.

If you did this, then the answer will depend as to what happens when this further delay is added to the programme, and the response will normally be, given the facts you have set out, that the date for planned Completion will be moved by 7 weeks. In that case, the Completion Date will be moved by 7 weeks (clause 63.5). This will be the same whichever way you dealt with the initial 3-week delay in your programme. And if you did not provide the programme for acceptance that you should have, then the *Project Manager* will not accept your assessment of the compensation event (third bullet of clause 64.1) and make its own assessment based on what it thinks the programme should be (clause 64.2), including, of course, the effect of the first 3-week delay.

Delays are therefore dealt with in the order that they are known about. The term 'dominant delay' is not mentioned in the ECC. Even with contracts that use such a term, the law would say that both are 'dominant' at different times, so the outcome would be the same. The *Project Manager* appears to be correct to assess this delay due to the compensation event as being 7 weeks, not 10 weeks.

137. Dealing with multiple programmes that have not been accepted (clause 63)

QUESTION

During the execution of an ECC Option C contract, programmes were submitted on a regular basis, and these were accepted by the *Project Manager*. However, the *Client* changed the *Project Manager*, and over the next 5 months no programmes were accepted. Most of the revised programmes were ignored and others not accepted for reasons not stated in the contract under clause 31.3. The contract does not appear to deal with this scenario other than, in clause 60.1(6), to state that no reply by the *Project Manager* within the period required is a compensation event.

In order to agree the effect that each compensation event had on planned Completion, a delay analysis has been produced, by working from the last Accepted Programme, inputting the time impact of the 'critical path' event, to establish a revised programme and planned Completion. This new programme (although not accepted) is then used to assess the delay of the next compensation event, and so on, forming a step-by-step process and a series of non-accepted programmes with the delays clearly identified and the new planned Completion demonstrated. These delays are identified within the appropriate compensation event quotation. Although this situation should not have arisen, can you advise how the contract envisages such a situation will be dealt with?

ANSWER

If the *Project Manager* fails to reply within 2 weeks to the programme, that will be a compensation event under clause 60.1(6). Equally, if the *Project Manager* does not accept the programme for a reason not stated in the contract, that will again be a compensation event, this time under clause 60.1(9).

So, the contract deals with both scenarios, and you should have notified these compensation events at the time and in accordance with the contract. If there is a dispute between you as to what the Accepted Programme was or should have been, then the *Adjudicator* will have to decide that. The *Adjudicator* may well start from the premise that the programmes you submitted at the time were correct, as they were not rejected for reasons within the contract by the *Project Manager*. It is important to understand that clauses 63.1 and 63.5 require that the assessment is made using forecasts made with the foresight the Parties had at the time, not the hindsight they have now. So, you use the Accepted Programme to assess the compensation event, then revise the Accepted Programme (at the intervals in the contract) to consider all of the delays up until then, including your own, as well as the compensation events, to come up with what the Accepted Programme should then have looked like, and so on. So, it is not just a case of feeding the delays caused by compensation events in each month; you need to allow for your own delays and the like. That is our view as to how the matter should be dealt with. However, it does not follow that every *Adjudicator* would agree, and for that reason we are unable to offer detailed advice and suggest that you should get further legal advice if you are still unsure.

138. Which programme to use when assessing compensation events (clause 63.5)

QUESTION

I am a *Contractor* assessing a substantial compensation event. I am confused as to which programme I should be using to assess a compensation event and the methodology. Is it the latest Accepted Programme or the one relevant at the time of the compensation event?

ANSWER

Clause 63.5 is clear on which programme to use – it is the Accepted Programme current at the 'dividing date'. The 'dividing date' is detailed in clause 63.1.

139. Including risk in compensation events (clause 63.8)

QUESTION

I am a *Project Manager* on an ECC Option B project. My question is about compensation events and whether risk should be included in the *Contractor's* quotation? If so, where does it say this in the contract?

ANSWER

Under clause 63.8, both time and cost allowances need to be assessed for risk within each compensation event. The clause states that they need to have a significant chance of occurring. We tend to find that if the *Client* has amended the compensation event section by way of Z clauses, then it may get a nasty surprise here, as the compensation events, quite rightly, will be more expensive and include more time risk allowance (as the *Contractor's* risks are greater).

Pouring a concrete slab in the UK in December in an exposed location would naturally include time and cost risk allowance to account for the impact of bad weather. If the concrete pour was internal/protected, then this would not be appropriate. This aspect is best discussed between the *Project Manager* and *Contractor* in order that they can understand each other's logic and thought processes and ensure that compensation events are managed in a timely fashion. The way we look at this is that everybody has a duty to their respective organisations to manage professionally and inform them of the time and cost impact of compensation events.

140. Inconsistencies in the *Contractor's* Scope (clause 63.10)

QUESTION

On our hospital project under ECC Option C, all the design responsibility is with the *Contractor*. We have signed the contract, but since discovered that its building drawings contradict the mechanical and electrical details in several instances. For example, one drawing for a particular room shows three light fittings whereas another shows five. We want five, and the *Contractor* is telling us that this would be a compensation event. Where do we stand?

ANSWER

The answer is quite simple and is contained in clause 63.10. Presumably, in the Scope you have clearly stated which aspect of the design the *Contractor* is responsible for (as per clause 21.1). We therefore have an inconsistency – this should be notified under clause 17.1. The *Contractor* should then be instructed to correct its Scope (clause 14.3) in order to correct the inconsistency.

The answer to whom the inconsistency is found most favourable to is contained in clause 63.10; it is found most favourable to the Party that did not provide the Scope – you. So, it is not a compensation event, as the *Contractor* proposed three and elsewhere stated five pendants (its inconsistency); if you want five, then there is no time and cost impact. If you wanted three, then it would be a negative compensation event.

This also works the other way, because any inconsistency or ambiguity in the *Client's* Scope is found most favourable to the *Contractor*.

The moral of the story is therefore (before contract signature) to ensure that each other's Scope is free of ambiguity and inconsistency.

141. Unclear Scope (clause 63.10)

QUESTION

An ECC Option A contract was tendered in which there were several reinforced concrete headwalls. There were no reinforced steel bar bending schedules accompanying the tender documents – a small reference box was included on the reinforcement tender drawings stating 'estimated reinforcement quantities for structures shown on this drawing for tender purposes only – 164.71 tonnes'. The contract was awarded in September with a starting date of November. The contract documents state that the steel bar bending schedules are included in the contract and will be provided on award of the contract; but the contract drawings still contain the reference box with the estimated quantities.

The *Project Manager* changed the Scope and notified a compensation event in February, stating: 'Estimated steel quantities were used at the tender stage of the contract, but the actual amount of reinforcing required is different from the tendered amount.' The net effect is that the amount of reinforcing has been reduced by 38.21 tonnes.

Would estimated quantities on a tender drawing be deemed to be in the Scope?

ANSWER

The tender drawings should be the same as the contract drawings. If you are later asked to enter a contract based on a Scope (including, of course, the drawings) that is different from that which your tender *activity schedule* is based on, you should refuse, as that could be a major problem! If you are not careful, the formal contract will be whatever you agree to at the time, and you might be stuck with a different Scope and *activity schedule*. However, for the reasons below, this does not appear to be the case here. The answer to your question is yes – as it is the contract Scope that you will have assumed was considered when pricing the work. So, what exactly was in the Scope in the contract – did it include both the drawings and the bending schedules, or were the bending schedules issued afterwards? If the answer is that both were in the contract Scope, then that is an inconsistency that the *Project Manager* should issue an instruction to resolve (clause 17.1). Presumably, the *Project Manager* instructed you to use the bending schedules? If so, that was a change to the Scope, in the sense that the *Project Manager* instructed you to ignore the drawings. That is a compensation event under clause 60.1(1). And that compensation event will be assessed using the principle set out in clause 63.10. As the Scope was provided by the *Client*, then it is assessed in your favour: that is, you are assumed to have allowed for in your price the lower quantities set out in the bending schedules. There is therefore no reduction, and the compensation event has no effect on cost or time.

On the other hand, if the bending schedules were not in the contract Scope, then your price is assumed to be based on what is on the drawings – after all there is no inconsistency, is there? When the *Project Manager* subsequently issues the revised schedules, that will be a change to the Scope, which is again a compensation event under clause 60.1(1). It will be assessed using the principles in clause 63.1 – that is, in general terms, how much did the change affect the Defined Cost for carrying out the work plus the Fee? And if that is a negative figure (as it is in this case), the total of the Prices is reduced (clauses 63.3 and 63.4).

142. *Contractor* not assessing compensation events (clause 64.1)

QUESTION

We are becoming very frustrated on our ECC Option C project as the *Contractor* is not assessing compensation events within the 3-week period. In some instances, we have extended the timescales under clause 62.5 in order to give the *Contractor* more time, but still, it fails to assess.

The *Client* is increasingly annoyed as it wants to know where it stands. What provisions are in the contract for dealing with this?

ANSWER

The contract is clear on how this is dealt with. Clause 64.1 provides four reasons/scenarios in which the *Project Manager* must step in and make its own assessment of a compensation event.

Note that the opening sentence states that 'the *Project Manager* assesses' – it does not say 'may' (which is discretionary). So, you must step in and make your own assessment – this is your obligation in the contract. It is very important to note that you need to act as stated in the contract (clause 10.1), and act in a spirit of mutual trust and co-operation (clause 10.2). Even if it is now up to you to assess, we would still be seeking to meet with the *Contractor* and ensure that the assessment is accurate. You also need to take account of other assessment clauses in the contract.

143. When is a compensation event 'implemented'? (clause 66.1)

QUESTION

When does a compensation event become agreed/accepted on an ECC project?

ANSWER

There is a clearly defined process for managing compensation events under the ECC (i.e. change control). This consists of a four-stage process: notify, quote, assess, implement. It is important to use the correct terminology – we do not agree/accept compensation event assessments as such. A compensation event achieves the status of having been implemented based on three potential reasons, as set out in clause 66.1: the *Project Manager* accepts the *Contractor's* quotation, the *Project Manager* notifies its own assessment or the *Contractor's* quotation is treated as having been accepted (refer to clause 62.6).

If any of those three scenarios occurs, then it has achieved the status of being 'implemented'. At that point the revised Prices and Completion Date will apply.

144. What is an implemented compensation event? (clause 66.1)

QUESTION

Can you define what an implemented compensation event is under the ECC? For example, does it mean when it is acknowledged by the *Project Manager* as the *Client's* liability, or when a quotation has been agreed? Would it be reasonable to suggest that liability for a compensation event could/should be agreed at ongoing early warning meetings, allowing the programme to be updated regularly and remain meaningful, even though the compensation event may not have been priced? If you agree with this approach, how should this agreement be recorded ideally?

ANSWER

The way that compensation events are implemented is set out in clause 66.1, and the consequences in clause 66.3. Early warnings and early warning meetings have nothing directly to do with compensation events. They are about matters that may occur (i.e. a risk) in the future, and are about managing that risk, whoever's it is. Compensation events are about things that have happened or will happen and are at the *Client's* risk because they are listed as a compensation event. Not all early warning notifications are about future compensation events, and not all compensation events will have an early warning.

We recommend that you keep the processes for dealing with risks of things that may happen separate from dealing with compensation events. And again, updating programmes is something different. The programme has to show both the Completion Date (contractual) and planned Completion (likely) (clause 31.2). The latter is linked through the activities in the programme, the former is not. The latter can move for all sorts of reasons, the former can only move because of a compensation event. If you know something is going to happen that you think is a compensation event, then its effects go on the programme straightaway and affect planned Completion.

145. Having your cake and eating it! (clause 66.3)

QUESTION

We have a *Project Manager* on our ECC Option C contract that, every time our cost is lower than the compensation event quotation, is assessing based on actual cost. Conversely, when our costs exceed our quotation, the *Project Manager* uses the quotation. The saying 'having your cake and eating it' springs to mind here.

What is the correct manner of assessing compensation events and which specific clauses do we need to refer to?

ANSWER

There are several clauses that define how compensation events should be assessed. Clause 63.1 states that this is largely a forecast of cost; clause 63.8 ensures that time and cost risk allowances are included (as appropriate); clause 63.9 also states that a compensation event is based on the assumption that the *Contractor* reacts competently and promptly, for example.

The key clause in this question is clause 66.3. This states that an implemented compensation event (both time and cost) is not revised except as stated in the *conditions of contract* (such as by the *tribunal*).

The compensation events process under the ECC ensures that a reasoned estimate becomes the basis of the compensation – not what the actual cost/time was. The actual figures are academic and not relevant – it is all about having a reasoned estimate and that becoming the basis of change to the Prices, Completion Date and Key Dates as appropriate.

146. Can you change an accepted quotation for a compensation event? (clause 66.3)

QUESTION

Given that the intent in the ECC is that a compensation event is quoted for, the quotation is agreed, and the work carried out, and the issue therefore closed, is there any provision for amending a quotation if it is found later to contain significant costs that should not have been accepted?

ANSWER

The first thing to say is that most compensation events should be assessed on a forecast of Defined Cost, not actual Defined Cost (see the last part of clause 63.1). With that in mind, the answer to your question is no, it cannot be revised. Once the *Project Manager* has accepted a quotation for a compensation event, it is implemented (clause 66.1). It cannot subsequently be revised if the forecast on which it is based is later shown to be wrong (clause 66.3).

147. Rolling up compensation events (clause 66)

QUESTION

We have just learned of a worrying trend in dealing with compensation events on an ECC contract. There have been a high number of compensation events that have been implemented. It appears that the *Project Manager* has agreed with the *Contractor* that rather than assess the time impact for each compensation event one compensation event at a time, all compensation events are assessed at the end of the month and assessed all together prior to submitting the clause 32.2 revised programme for acceptance. A separate compensation event is then notified if there is any change to the Completion Date.

It is now 6 months into the contract and the Completion Date has been extended by 6 weeks. The *Project Manager* was unfortunately taken ill, and our replacement *Project Manager* has advised us that in reviewing the contract progress so far, there is a concern about the agreement on assessing the time impact of compensation events. The *Project Manager* says that the *Contractor* is updating the percentage completion of activities after inputting the time impacts of compensation events. This results in an incorrect movement of planned Completion to the right, whereas if they were all input contemporaneously, there would be no change to planned Completion. The *Project Manager* has advised us that it is unable to change the revised Completion Date but will be reducing the Prices to remove the prolongation cost associated with some of the compensation events implemented to date.

The *Contractor* is also concerned and has contacted us to say that we should instruct the *Project Manager* to put things right or the matter will be referred to adjudication. What should we do?

ANSWER

Once the *Project Manager* has accepted a quotation for a compensation event, it is implemented (clause 66.1). It cannot subsequently be revised if the forecast on which it is based is later shown to be wrong (clause 66.3). But of course, the contract appears to have been largely abandoned here in terms of the requirements of dealing with compensation events.

Going forward, the replacement *Project Manager* should follow the contract provisions correctly, as required in clause 10.1. However, the problem here might be the changes to the Prices and any dates may well have been incorrectly made in the past, so there is a risk of compounding the error. If the outcome to date looks about right, it may be that the Parties could draw up an agreement to reflect this in accordance with clause 12.3. If either or both Parties believe what has happened is quite wrong and they cannot find some sort of agreement on this, then adjudication may be the only way forward to unravel this, assuming Option W1 or W2 is incorporated.

148. Extending time for the submission of a quotation for a compensation event (clause 62.5)

QUESTION

With the agreement of the *Contractor*, the *Project Manager* has extended the time allowed for the *Contractor* to submit quotations for ten large value compensation events. Again, with the agreement of the *Contractor*, the *Project Manager's* assistants have been discussing with the *Contractor* the detail of draft quotations for each compensation event. We were about to confirm agreement when the *Contractor's* senior surveyor was replaced. The first action by the *Contractor's* new senior surveyor was to suspend any discussion on the compensation events. A week later the *Contractor* formally withdrew the quotations. The revised date for submission of the quotations has now passed. What happens next?

ANSWER

Under clause 64.1, the *Project Manager* will be required to assess each compensation event because the *Contractor* has not submitted the quotation and details of its assessment within the time allowed. How long the *Project Manager* must do this is stated in clause 64.3.

149. Using existing Bill of Quantities rates (clause 63.1)

QUESTION

The *Contractor* wishes to use an existing rate in the Bill of Quantities in an Option B ECC to assess a compensation event. We feel it is considerably above what we expect the market rate is for such work. Do we have to use this rate for assessing a compensation event?

ANSWER

The default method of assessing compensation events in all main Options is clause 63.1. Effectively, this is a bottom-up quotation based on open market rates or competitively tendered prices in accordance with clause 52.1. If the *Contractor* and *Project Manager* agree, they can used existing rates or lump sums, which in turn can be taken from the Bill of Quantities (clause 63.2). That seems unlikely in this case and the default clause 63.1 should be used.

150. The *Project Manager's* assessments (clause 64.1)

QUESTION

Can you explain what happens if, when the *Contractor* submits a quotation for a compensation event, the *Contractor* has not submitted a revised programmed in accordance with clause 32.2?

ANSWER

The third bullet of clause 64.1 states that, where the *Contractor* has failed to submit a revised programme required under clause 32.2, then it falls to the *Project Manager* to assess a compensation event even if the *Contractor* has submitted a quotation.

151. Prevention (clause 60.1(19))

QUESTION

ECC clause 60.1(19) always seems to cause interesting discussions, with clients either omitting it by means of a Z clause, or more commonly amending it.

It is also tied into clause 19.1 prevention, following which and if the event occurs, the *Project Manager* instructs the *Contractor* how to deal with the event.

We accept that the NEC contracts do not include provisions for force majeure – that is, significant events such as major flooding, earthquakes, volcanic eruptions and the associated dust clouds, pandemics and other natural disasters that prevent the contracting parties from fulfilling their obligations under the contract. Clause 60.1(19) essentially frees the Parties, or gives them some relief from their liabilities. It can encompass minor issues and is open to interpretation.

For example, the Parties have deleted clause 60.1(13) from the contract by means of a Z clause. One would then assume that the risk of unforeseen weather conditions is the *Contractor's*, and it should have allowed for it within its tender. The Site is affected by a severe storm that occurs less frequently than once in 10 years, which delays Completion of the whole of the *works* by 1 week.

The *Contractor* alleges the following.

- The weather stopped it completing the whole of the *works* by the date shown on the Accepted Programme. Assume that the *Contractor* can prove that.
- Neither Party could prevent the event. While one can take measures to lessen the impact of a storm on the progress of the *works*, one cannot prevent the storm from occurring.
- An experienced contractor would have judged at the Contract Date to have such a small chance of occurring that it would have been unreasonable to have allowed for it. The *Contractor* states that it could not have predicted or allowed for such a rare event and its consequences.
- It is not one of the other compensation events. If clause 60.1(13) has been deleted from the contract by means of a Z clause, one could argue that it is no longer one of the other compensation events.

Can the *Contractor* notify a weather-related compensation event because the specific weather-related compensation event clause 60.1(13) has been omitted from the contract?

ANSWER

An interesting question and a lot to work through.

Let us start with the significant events such as pandemics, earthquakes and so on. The ECC should not be considered as a complete contract that will work in every sector in every corner of the world without some sort of adjustment. You will need to localise it for these types of natural disaster and consider adding some of them in, perhaps as *Client's* liabilities or compensation events. This may take out some of the angst in deciding whether a prevention event has occurred. Additional *Client's* liabilities and additional compensation events can be simply added through entries in the Contract Data.

In your example, the deletion of clause 60.1(13) means there is no specific compensation event for whatever weather comes the *Contractor's* way. Somehow, the *Contractor* will have to factor this risk into their bid. Should there be compensation events that occur once the Parties are in contract, then the *Contractor* may be able to make due allowance in their quotations for future weather that has a significant chance of occurring. The weather risk sits with the *Contractor* for Providing the Works but where, for example, the Scope is changed by the *Project Manager's* instruction, the effect on the *Contractor's* risk should be considered further in the quotation for the compensation event. Should quite inclement weather occur while carrying out any additional works, there is still no means in the contract to further adjust the implemented compensation event as there is no compensation event covering weather events.

Later in your example you suggest the critical path is delayed by a week due to the inclement weather. The ECC does not use the term inclement weather by the way, it is just a term we are using to explain things. To delay may not be enough here. The second bullet of 60.1(19) is looking for events that 'stop the *Contractor* completing the whole of the *works* by the date for planned Completion shown on the Accepted Programme'. What we do not have here is evidence there is an Accepted Programme nor evidence that no matter what the *Contractor* does, it is stopped from achieving Completion by the date shown on the Accepted Programme for planned Completion. Could it re-sequence, introduce additional resources and so on? Let us assume not; no matter what the *Contractor* does, it is stopped. So, we are passing that part of the test, and you did say assume it could prove this, but that is quite a high threshold to pass normally.

We agree that neither Party could prevent the event and it is not one of the other compensation events. So, we are left with the experienced contractor/small chance of occurring test. Surely this depends on the event and now there seems to be a radical shift in the frequency of these so-called rare events. While we wait for statistics to catch up and adjust, it appears what was a 1 in 10 is looking more like 1 in 1 these days. Is anyone assuming when bidding right now that the weather will be the same as what it might have been 10 or 20 years ago? These are tricky discussions and there is clearly some subjectivity about it, but it could well be that certain weather events could fit into this category of compensation events. The consequence may not be relevant, it is the event itself that we are judging and what does 'rare' mean? That is not the language of the contract.

As an aside, extreme weather events may not be the sorts of events envisaged when deciding which *weather measurements* to include in the Contract Data when tendering. The best way to avoid such uncertainty will be of course to properly think about appropriate and clear risk allocation when preparing the contract in the first place.

152. The time bar in clause 61.3 (clause 61.3)

QUESTION

Does the time bar in clause 61.3 automatically kick in, or must the *Project Manager* or *Client* do something first?

ANSWER

It is in the hands of the *Contractor* to choose to notify in good time, or not. Clause 61.3 explains by when certain compensation events must be notified and what happens if they are not. No action is needed by either the *Project Manager* or *Client*.

153. The dividing date (clause 63.1)

QUESTION

Why isn't the dividing date a defined term?

ANSWER

Well, it is not in the current drafting but may be better drafted as a defined term in any future editions. As defined terms are almost all in alphabetical order, to add something like this has a considerable impact in re-numbering a significant number of references and clauses in NEC4 contracts, User Guides and flow charts!

FURTHER THOUGHTS AND TOP TIPS

- Compensation events do not manage themselves – agree at the project start on how these will be managed. Consider a fortnightly meeting to keep on top of this.
- Follow the process in the contract. Do not skip stages.
- Include the planner – both cost and time need to be assessed jointly. Time is money!
- Remember clause 63.1 – compensation events are largely a forecast. Time and cost risk may be included.
- Use the ability to make *Project Manager* assumptions under clause 61.6. This is an excellent means of allocating risk and obtaining better value for money.

Robert Alan Gerrard and Patrick Waterhouse
ISBN 978-1-83708-959-8
https://doi.org/10.1108/978-1-83708-956-720252007

Chapter 7
Title

154. Does title mean ownership? (clause 70)

QUESTION

We have been advised that the Engineering and Construction Contract (ECC) does not contain entitlement to payment for Plant and Materials on or off Site. We note that clause 70.2 states that title passes to the *Client* for Plant and Materials brought within the Working Areas. Does 'title' mean 'ownership' and, if so, how can this pass if no payment is made?

ANSWER

Things are never quite as simple as they seem, and 'title' is not quite the same as 'ownership'. Payment and title do not have to be directly linked, as title can pass without payment being made, and payment can be made without title passing. It will all depend on the terms of the contract that the Parties have entered into and the *law of the contract*. In the case of the ECC, title can pass without payment being made. It is, by the way, the same with most standard form construction contracts, as they will generally state that title will pass as soon as such material arrives on site. You will often find, though, that payment is rarely made in advance or at that time and will often be made some time afterwards. Even then it will often only be at 80% or so until the materials are incorporated into the *works*.

For the ECC, it is also not the case to say that in all Options the *Contractor* will not be paid for Plant and Materials on Site. In Option A, it will depend on how the Activity Schedule has been completed. For example, if there is an activity that says 'Item x delivered to Site', then the *Contractor* will be paid that amount in the assessment after item x arrives. Equally, if an item in the Activity Schedule says 'Complete manufacture of item x', then when it is complete the *Contractor* will be paid in the next assessment even if the Plant and Materials are not on Site. See clauses 70 and 71 for the safeguards in that case. With Options C, D, E and F, the *Contractor* will often be paid Defined Cost before it pays for such Plant and Materials, and that will include for Plant and Materials wherever they are (clauses 11.2(31) and 11.2(24) and item 3 of the Schedule of Cost Components (SCC)).

155. Vesting Plant and Materials (clause 70.2)

QUESTION

On an ECC Option A contract, does the inclusion of clause 70.2 avoid the need for vesting documentation for Plant and Materials on Site if the area for storage is included in or added to the Working Areas under clause 16.3?

ANSWER

We are limited in what we can say because the question of vesting certificates is a legal question. Once any Plant and Materials (not Equipment) are brought within the Working Areas, whatever title the *Contractor* has to them passes to the *Client* (clause 70.2). However, the law around this is always complex and dependent on the *law of the contract*, and, as a rule, the *Contractor* cannot pass to you title it does not have, whether by the operation of clause 70.2 or by the signing by the *Contractor* (as opposed to the title holder) of a vesting certificate.

Regarding storage, it is important to understand that the definition of the Working Areas can only include areas that are being used solely for work on the contract (second bullet of clause 11.2(20) and also clause 16.3). Therefore, the area will and should be under the direct control of the *Contractor*. The more remote from the *works* the Working Areas are, the more risk there is to the *Contractor* and *Client* when it comes to Plant and Materials being removed for whatever reason. This is a risk that the *Client* will need to manage.

156. Title to materials from excavation (clause 73.2)

QUESTION

On an ECC Option D contract, who has title to surplus materials from any excavation works?

ANSWER

Clause 73.2 states that the *Contractor* has title to materials from excavation and demolition unless the Scope states otherwise. The default, therefore, is that it will be the *Contractor* that has title to such materials unless stated otherwise in the Scope. This will, of course, demand careful thought at the tender stage to decide what is best for a particular project. But you should note that any sale of these materials to a third party would require a credit under paragraph 54 of the Schedule of Cost Components.

FURTHER THOUGHTS AND TOP TIPS

- Make sure when drafting the ECC Scope that it contains an appropriate statement if you wish the *Client* to have title to materials from excavation and demolition.

Robert Alan Gerrard and Patrick Waterhouse
ISBN 978-1-83708-959-8
https://doi.org/10.1108/978-1-83708-956-720252008

Chapter 8
Liabilities and insurance

157. Weather ... it's a compensation event (clause 80.1)

QUESTION

I am trying to understand risk under the Engineering and Construction Contract (ECC). Clause 80 seems to be quite clear as to who bears liability for what. If the *Client* does not amend Contract Data part one to add to the standard *Client's* liabilities under clause 80.1, why for example does an abnormal weather event become a compensation event under clause 60.1(13)? Surely, if weather is not included in Contract Data part one, then it is the *Contractor's* liability – full stop? So, how can there be compensation?

ANSWER

It is important to understand that the liabilities in clause 80.1 merely cover things such as liabilities to physical damage and third-party responsibilities. They do not cover the financial risks for carrying out the work itself. That financial (and time) risk is set firstly by the list of compensation events. Compensation events are always at the *Client's* risk. The other major factor regarding the remaining financial risks (although not time risks) for carrying out the *works* is the choice of the main Option. With Options A and B, most of the financial risks for carrying out the *works* lie with the *Contractor*. In Options C and D, most of them are shared between the Parties. In Options E and F, most of them lie with the *Client*.

158. A hypothetical situation (clause 80.1)

QUESTION

Clause 80.1 states

> The following are *Client's* liabilities. Claims … and costs payable to Others which are due to … use or occupation of the Site by the *works* or for the purpose of the *works* which is the unavoidable result of the *works*.

Would it be correct to say that in a hypothetical situation, if a contract required a large number of heavy goods vehicle movements to import a very large volume of stone, and if the 3.5 m access road that formed the access to the Site, and was within the limits of the Site, started breaking up, that the costs associated with remediating the access road would fall within the coverage of the stated part of clause 80.1 and hence be the *Client's* liability?

ANSWER

Nice try but … no! 'Unavoidable' means that it could not be avoided no matter what you did, and we cannot see how this damage could be called 'unavoidable'. You could have avoided it by carrying out the stone placing in a different way – using smaller lorries, or dumpers and stockpiles. Or you could have put a protective mat over the road, or you could have constructed a different temporary road and used that instead. All of those would have avoided this damage, so it cannot be said to be 'unavoidable'. All would cost much more, but that is irrelevant. 'Unavoidable' does not mean the same as 'not economically viable'.

159. Copper theft! (clause 80.1)

QUESTION

We have been employed by our client (the main contractor), which has decided that the contract we are working under is ECS Option A – which is back-to-back with the *Client*. The question is that when working on Site we had to leave materials (copper fittings) in the building, and I was under the understanding that the Site should be safe and secure. There was a theft, and I would like to know who is responsible for this?

ANSWER

The first point we need to make is that NEC contracts cannot be entered into just by reference to them. There is an important document called the Contract Data that needs to be completed. In your case, assuming you are using the ECS, the *Contractor* (your client) needs to have filled out Contract Data part one, and you need to have filled out Contract Data part two. Without that information the 'contract' is simply too uncertain to be valid. If you do have a valid ECS contract, then the answer is normally that you are entirely responsible for the security of the Plant and Materials (a defined term) left on Site, and you must insure against their loss (see the Insurance Table in clause 83.3). Under the ECS, this is thus your liability (clauses 80.1 and 81.1). You therefore have to look to your insurers to recover these costs. The only possible exception is that if the theft was as a result of the fault of the *Contractor* (your client) or the *Client* (the *Contractor's* client) – for example, if either had left the Site unlocked at night when it should not have – then that may be the *Contractor's* liability (second main bullet of clause 80.1). If that is the case, then it would be a compensation event under clause 60.1(14) of the contract. Compensation events are the only way that you will get paid any more money or be allowed any more time under the ECS. The process involved is set out in clauses 60 to 66. It is important for you to understand the processes involved – for example, you must notify a compensation event within 7 weeks of becoming aware of the event, otherwise you lose your entitlement (last sentence of clause 61.3).

160. Is terrorism a compensation event? (clause 80.1)

QUESTION

Terrorism is not referenced as a *Client's* liability within the ECC under section 8. Does this mean that, by default, acts of terrorism are a *Contractor's* liability and something that should be included in the *Contractor's* insurance policy? Or should the contract be specific and ask the *Contractor* to take out specific insurance cover for acts of terrorism? If no specific request is made, then will it be a *Client's* liability and treated as a compensation event under clause 60.1(14) if an act of terrorism occurs?

ANSWER

As you rightly say, the answer to this will be in section 8 of the contract. You are also right to say that terrorism is not specifically mentioned in the list of *Client's* liabilities in clause 80.1. However, there are items such as civil war, rebellion, revolution, insurrection, riots and civil commotion. Whether these cover any or all acts of terrorism is a legal question. We can only suggest that you get some specialist legal advice on this. If this is covered, then it is not your liability and you do not need to insure against it. However, if terrorism is not included within clause 80.1, then clause 81.1 of the contract clearly makes it your liability. You will therefore be required to provide against it for the insurances set out in the first two rows of the Insurance Table.

Many national governments have agreements in place with the insurance markets to fund payments resulting from terrorism. The answer we have given here can only explain the position under unamended NEC contracts. Many NEC contracts are amended to reflect specific insurable risks, such as terrorism.

161. Take over versus Completion (clause 80.1)

QUESTION

I am confused by the terms 'take over' and 'Completion' under the ECC. We have a situation in which the *Contractor* achieved Completion 1 week prior to the Completion Date. Completion was certified but a member of the public damaged the *works*. Who is liable?

ANSWER

The first thing to remember is that take over and Completion are two different concepts and are not directly related. Completion is about the state that the *works* are in (clause 11.2(2)), whereas take over is about who becomes responsible for the *works* (clause 80.1). As you rightly say, the *Contractor* remains responsible for the *works* (and must insure them) until take over, not Completion, has happened. And take over can occur before, at or after Completion. To add to the mix, you also need to remember that the Completion Date is the date by which the *Contractor* is obliged to achieve Completion (clause 30.1). And again, Completion can be achieved before, at or after the Completion Date. Of course, if it is the last of those three, the *Contractor* will be in breach of contract.

The 'normal' position is that take over occurs within the 2 weeks after Completion, whether the *Client* wants to or not (second sentence of clause 35.1). So, if the *Contractor* achieves Completion 6 weeks before the Completion Date (i.e. 6 weeks 'early'), then take over occurs 2 weeks afterwards (i.e. 4 weeks before the Completion Date). However, for some projects this may not be convenient for the *Client* – for example, its staff may not be ready to occupy it early. In that case, the *Client* may, before making the contract, make it clear to the *Contractor* that it does not want to take over the works before the Completion Date. That is done by inserting an optional entry in Contract Data part one. So, the *Contractor* then knows that when it is pricing the work that if it finishes early, it will still be responsible for the *works* until the Completion Date, and can price that risk accordingly. In that case (and only in that case), the first sentence of clause 35.1 applies.

In addition to all of this, the *Client* may, if it wants to, start to use all or any part of the *works* at any time, which can be before, at or after Completion has been achieved (first sentence of clause 35.2). If it does so, it takes over those parts of the *works* (second sentence of clause 35.2), unless the exceptions listed in the bullets of clause 35.2 apply. So, if the *Client* starts to use the *works* as soon as Completion occurs, take over occurs at that point. Also, take over before Completion has been achieved will be a compensation event unless it happens after the Completion Date (clause 60.1(15)). Whenever take over occurs, the *Project Manager* is required to certify it (clause 35.3). Usually, Completion and take over occur on the same date, because the *Client* is keen to start using the *works*. In that case, the *Project Manager* issues two certificates for that day: one for Completion and one for take over.

In your case, assuming there is no optional statement in the Contract Data, if the *Client* had not started using the *works* and if the damage occurred within 2 weeks of Completion, then the *Contractor* is liable for this damage.

162. How to define take over (clause 80.1)

QUESTION

I work for a local authority, and we are procuring some highways repair work under the ECC.

I want to make it clear that the various take overs, which allow the public to use the highway we are working on, are not our liability. I want to avoid a situation in which damage caused by a car accident, for example, could be construed as a *Client's* risk. How do I clarify this in the contract documents?

ANSWER

The answer lies, as it often does, in careful drafting of the Scope. If you follow the ECC User Guides on writing the ECC Scope (which is advisable), then you are directed to the way this should be drafted. The guide draws attention back to clause 35.2 of the ECC. This states that the *Client* may use any part of the *works* before Completion has been certified. If so, take over begins when it begins to use it. However, the clause goes on to state 'except' (first bullet) for a reason stated in the Scope.

So, if you describe the reasons clearly within the Scope, then (as per your aim) the *Contractor* will remain responsible.

163. Insurance of existing buildings (clause 83.3)

QUESTION

I am currently drafting an ECC form of contract and would be grateful if you could advise whether insurance of existing buildings (to be retained and refurbished as part of the *works*) must be taken out by the *Client*, or is included within the insurance provisions contained in the Contract Data?

ANSWER

If you look at the Insurance Table in clause 83.3, the first row of the table covers insurance against loss of or damage to the *works*, Plant and Materials. This covers such loss and damage to the new *works* themselves.

The third row of the table details insurance against loss of or damage to property (except the *works*, Plant and Materials and Equipment). This covers, for example, property other than the *works* itself.

The *Contractor* is thus obliged to provide both insurances unless otherwise stated in Contract Data part one (clause 83.2).

It may be that the *Client* has an insurance policy in place that would cover the existing buildings and their refurbishment. If the *Client* has not got such insurance, then the default contractual position is *Contractor*-provided insurance as stated.

164. Damage caused during a search (clause 84.2)

QUESTION

Using the ECC, I understand that if the *Supervisor* asks the *Contractor* to search for a Defect and no Defect is found, then a compensation event arises. Conversely, if a Defect is found, the *Contractor* remedies this at its own cost (depending on the main Option chosen). What happens when the *Contractor* damages a part of the *works* and essentially creates a Defect while carrying out the search? Is the *Contractor* entitled to recover the remedial costs as a compensation event or can the *Contractor* argue that creating a Defect was a risk of carrying out the test?

ANSWER

The answer is going to depend on the circumstances. If the test or inspection made the damage inevitable, it should be paid for as part of the compensation event because it was not accidental damage. So, for example, if the *Contractor* had to break out the newly constructed car park surface to test or inspect something underneath, then breaking it out and reinstating it is all part of the Defined Cost for carrying out the test or inspection.

On the other hand, if the damage was accidental in the sense that it could have been avoided but was not, then the *Contractor* is required to insure against such damage (see the first row in the Insurance Table in clause 83.3). The *Contractor* will have to look to its insurer to cover such costs, and, if they are below the policy excess, that is the *Contractor's* liability.

165. Accepting insurance policies (clause 85)

QUESTION

I am a *Project Manager* on an Option C contract. I want to ensure that I do everything correctly but am concerned about my liability if I 'accept' a submitted insurance policy by the *Contractor*. Clause 84.1 requires that I 'accept' the policy, but I am no expert in the insurance field. Where do I stand with this?

ANSWER

The easy things to check for are that the amounts are compliant with the Insurance Table in Contract Data part one. You can also check that the certificates are signed by the *Contractor's* insurer or insurance broker (clause 84.1). However, reference to clause 14.1 should put your mind at rest, as your 'acceptance' does not change liability. Finally, it is recommended that communication is undertaken in accordance with the published ECC communication forms in volume 4 of the ECC User Guides. This will ensure a clear audit trail and record. You should not hesitate to contact the *Client* for further advice if you think that this task exceeds either your own or your company's expertise.

166. Failing to provide insurance certificates (clause 84)

QUESTION

What happens if the *Contractor* does not provide insurance certificates as required by the ECC, and what is the remedy?

ANSWER

The ECC requires (clause 84.1), before the *starting date* and on renewal of the insurance policy, the *Contractor* to submit certificates to the *Project Manager* for acceptance. If it does not, then the *Client* may take out these insurances, and the cost is paid by the *Contractor* (clause 85.1).

This also works the other way – the *Project Manager* is required to submit the same for risks that the *Client* is required to insure against. If it does not, then the *Contractor* can take out the insurance, and the cost is paid by the *Client*.

167. The difference between liabilities and compensation events (clauses 60 and 80)

QUESTION

A *Client's* liability is also stated as being a compensation event. So, what is the difference between liabilities and compensation events?

ANSWER

Various events will occur on a contract from time to time. We have the more likely events such as a change to the Scope, vandalism, theft, bad weather and unforeseen ground conditions, but also more unlikely ones such as force majeure, fire, war, terrorism and finding historical artefacts. If they occur, some of these may just lead to the *Contractor* incurring more cost and suffering programme delays, while others could lead to loss or damage to the *works* already carried out. The first thing the drafters do is decide which of these are the risk of the *Client*, and which are the risk of the *Contractor*.

Those things that are the risk of the *Contractor* and that cause things like loss or damage are stated as being *Contractor's* liabilities (clause 81.1), for which the *Contractor* must procure insurance to cover.

Those things that are the risk of the *Client* and that cause things like loss or damage are stated as being *Client's* liabilities (clause 80.1), for which the *Client* may procure insurance to cover if it wishes to. These also double up as being compensation events (see clause 11.2(14)) as they may, in turn, affect the *Contractor's* Defined Cost and/or the various dates in the contract.

Those things that usually have time and/or cost effects that the drafters decide are to be the *Client's* risks are called compensation events.

FURTHER THOUGHTS AND TOP TIPS

- Make sure that the Insurance Table and amounts are appropriately completed at the tender stage.
- Seek professional advice if at all unsure.
- Ensure that insurance details are provided to both Parties. Make a note of the renewal dates.
- Adding additional *Client's* liabilities is an excellent way of allocating risk fairly and obtaining better value for money.

NEC4: Practical Solutions, Second edition

Robert Alan Gerrard and Patrick Waterhouse
ISBN 978-1-83708-959-8
https://doi.org/10.1108/978-1-83708-956-720252009

Chapter 9
Termination

168. Is there a standard Engineering and Construction Contract (ECC) termination certificate? (clause 90)

QUESTION

Is there a standard ECC termination certificate template available?

ANSWER

The NEC does not provide a termination certificate template, as the provided templates are only for the most used actions under a contract. Termination rarely happens, and if it does, we strongly recommend that you take legal advice to ensure that the contract requirements are followed. It is not something that should be done lightly.

The request for a certificate is issued by one of the Parties. The certificate will be issued by the *Project Manager* if it is satisfied that the reason stated in the request applies and it complies with the contract. The certificate does not need to be any particular format; it just needs to comply with clauses 90.1, 13.1, 13.6 and 13.7.

169. Is there a standard Term Service Short Contract (TSSC) termination certificate? (clause 90.1)

QUESTION

We are using the TSSC and are looking to terminate in accordance with the provisions of clause 90.1 (for reasons 2 and 3 stated in clause 90.3). The clause suggests that we must issue a termination certificate. Can you advise whether there is a standard form for this, or whether it is anticipated that this will be a formal letter headed ‘termination certificate’?

ANSWER

There is no standard form for this. The communication must be 'received in a form that can be read, copied and recorded' (clause 13.1) and must make it clear what it is.

Note that this is a two-part process. Firstly, the *Client* needs to notify the *Contractor* that it wishes to terminate, and then it issues a certificate if the reason given complies with the contract. You must also bear in mind with reasons 2 and 3 of clause 90.3 that you must first give a notification of the default and then allow the *Contractor* 2 weeks to stop defaulting before you can notify your intention to terminate.

170. Can you terminate due to poor performance? (clause 91)

QUESTION

We are looking to terminate an ECC Option C contract due to several reasons: the main ones being poor performance in relation to programme and quality. How do we put this into effect?

ANSWER

Termination in the ECC terminates the *Contractor's* further employment under the contract, rather than terminating the contract itself (clause 90.1). In the wording of clause 90.1, you are terminating 'the *Contractor's* obligation to Provide the Works'. The procedures for termination and the amounts to be paid on termination are set out in clauses 90 to 93. Note that the procedures and amounts will depend on who is terminating and for what reason (see the Termination Table in clause 90.2). Assuming that you are the *Client* here, it is essential that you provide a valid reason for termination (clause 91) and then scrupulously follow the procedures set out in the contract. We suspect that clauses 91.2 or 91.3 would be the most relevant to you, but in that case, you need to notify the *Contractor* of the default and then allow it 4 weeks for correction before terminating.

We are unable to advise you whether the particular 'poor performance' would meet the requirements of being 'substantial' required by these clauses, because it is in fact sensitive and will be a matter of law. You should therefore take legal advice on that point before proceeding.

171. Can we use R21 termination for a minor earthquake? (clause 91.7)

QUESTION

On an ECC contract, if we suffer a minor earthquake during construction of the *works*, can we as *Contractor* terminate under clause 91.7 (R21)?

ANSWER

The simple answer is no, this right is only given to the *Client* as stated in the first line of clause 91.7. In any case, there are a few hurdles to clear for this reason to be valid – one of the first two bullets in clause 91.7 must be satisfied along with both last two bullets. The operative word in the first two bullets is 'stops'. The event that has occurred has to have stopped the *Contractor* completing the *works* (basically in its tracks, first bullet) or completing the whole of the *works* by the date for planned Completion shown on the Accepted Programme and be forecast to delay Completion by more than 13 weeks (second bullet). The clause goes on to test that neither Party could have prevented the event and finally that an experienced contractor (not the *Contractor*) would have judged it at the Contract Date to have such a small chance of occurring that it would have been unreasonable to have allowed for it. When broken down, you can see there are quite a few hurdles to clear in clause 91.7 before the *Client* could exercise the right to terminate.

172. Can you terminate for corruption? (clause 91.8)

QUESTION

When using ECC Option C, could you confirm that the *Client* can terminate for a corrupt act carried out by the *Contractor*?

ANSWER

Yes, that is exactly what clause 91.8 provides for. You just need to check that it is a Corrupt Act by the definition in clause 11.2(5).

FURTHER THOUGHTS AND TOP TIPS

- People will often refer to 'terminating the contract', whereas in fact under the ECC they are 'terminating the *Contractor's* obligation to Provide the Works'. There is a difference.
- Termination is a very specialist area, and great care should be taken to ensure that the exact provisions of the contract are followed and, in most cases, appropriate advice is obtained.

Robert Alan Gerrard and Patrick Waterhouse
ISBN 978-1-83708-959-8
https://doi.org/10.1108/978-1-83708-956-720252010

Chapter 10
The Options

173. Do I use Option W1 or W2? (Options W1 and W2)

QUESTION

I am completing the Contract Data for a new-build schools project in the UK, and we are using Option A of the Engineering and Construction Contract (ECC). My question is, which dispute resolution procedure do I use?

ANSWER

As you are in the UK, the work you describe will be covered by the Housing Grants, Construction and Regeneration Act 1996 and the equivalent legislation in Northern Ireland. This will mean that you will need to use Option W2 – this has been specifically written to comply with the legislation. You will also need to use Option Y(UK)2 in respect of payments. If you are in Northern Ireland, you should also use Y(NI)1 with Y(UK)2. Elsewhere in the world, or for projects in the UK that are not subject to the construction contracts legislation, you could use W1 or W3. But you should check local legislation first.

174. Option W1 or W2 – which applies? (Options W1 and W2)

QUESTION

I am involved in the construction of a major drainage project in Hong Kong. I am a bit confused as to which dispute resolution procedure applies – is it Option W1 or W2?

ANSWER

You could use either. Option W2 has been specifically drafted to comply with the UK's Housing Grants, Construction and Regeneration Act 1996 and its equivalent in Northern Ireland. The main difference between Options W1 and W2 is that Option W1 contains time limits for the notification of a dispute. Many users prefer this option, as you know where you stand in many respects.

175. Subcontracting and inflation (Option X1)

QUESTION

We are the *Client* under an ECC Option C contract incorporating secondary Option X1 price adjustment for inflation. The *Contractor* has placed its subcontracts under its own bespoke form of subcontract, which does not flow down from the NEC contract we have with the *Contractor*. All the subcontracts let are fixed-price lump sum plus any variations, and there is no mention of any escalation applying.

The *Contractor* is not paying any escalation amounts within its interim or final payments to its Subcontractors. The Defined Cost of Subcontractors is therefore exclusive of any escalation amounts. The question is, when assessing the *Contractor's* price adjustment factor amounts under X1, should the amounts for these Subcontractors not be included within the Price for Work Done to Date?

ANSWER

The amounts should be included, because the Price for Work Done to Date includes the costs of these Subcontractors (clauses 11.2(31) and 11.2(24)). Of course, the answer you come up with for the price adjustment is not added to the Price for Work Done to Date, instead it is added to the total of the Prices (i.e. the 'target') (clause X1.4).

Option X1 is just a convenient way of calculating an assessment of what the inflation may be, but there are all sorts of ways that the actual inflation can vary from that assessment. However, the risks, or rewards, of those variations are shared between the Parties through the share mechanism.

Just because there are no inflationary increases within a particular subcontract, it does not mean that the *Contractor* will not be paying for inflation. In that case, it could well be (and probably is) the case that there is an allowance for inflation within the Subcontractor's prices.

176. Indices calculations (Option X1)

QUESTION

We have an Option B ECC contract with Option X1 price adjustment for inflation. I am confused by clause X1.5 – why do we need to alter the compensation event assessment to the base date?

ANSWER

At first read this is slightly confusing; however, the logic here is that each compensation event is adjusted back to the base date (i.e. at tender). Clause X1.3 then adjusts the amount due by the indices calculation. If compensation events were not adjusted in this way, then the inflation would be double counted.

177. Adjusting for the price adjustment (clause X1.5)

QUESTION

We are using X1 on our ECC Option C contract. If I am reading clause X1.4 correctly, it seems to be calculating the inflationary amount based on cost, and not the Prices as you would under Option A or B. Am I correct in this interpretation, and what is the logic behind this?

ANSWER

Yes, you are correct. The calculation under Options C and D uses the Price for Work Done to Date. Under these Options this is Defined Cost (the Schedule of Cost Components (SCC)). Although it is almost impossible to assess correctly the actual amount of inflation that has been incurred on a project, this calculation attempts to be more realistic and derive the inflationary amount from the cost incurred as opposed to the Prices. In turn, this derived assessment is added to the Prices (clause X1.4).

178. I think I've found a drafting error in the ECC! (Option X2)

QUESTION

I am completing the Contract Data on our ECC project but there is no detail for Option X2 – is this a drafting error in the contract?

ANSWER

Drafting error in the ECC? No such thing! Option X2 is simply in or out, so no additional information is needed to complete Contract Data part one. The opening paragraph of Contract Data part one will determine whether this Option is used or not.

179. Is this a change in the law? (Option X2)

QUESTION

With reference to changes in national insurance (a form of taxation) in the UK, the *Contractor* has notified a compensation event under Option X2. I would appreciate a commentary on admissibility in the light of the facts that the relevant statutes were known of at the time of tender. Also, can this be a negative adjustment?

ANSWER

Regarding Option X2 the position is clear – it is a compensation event if the law changes after the Contract Date. It is irrelevant whether that change of law had been announced many months or years before. So, the question is, when did the law change – not when was the change announced? That means that the *Contractor* does not need to allow for the effect of any change in the law that has been announced but not yet enacted. On the other hand, if the law itself had changed before the Contract Date but was not to take effect until a later date, then that will not be a compensation event. We do not know the exact arrangements for bringing in these changed national insurance contributions, so you will need to get expert advice on that. However, if the effect of this law change is to reduce employers' payments, then that needs to be considered in the assessment of the compensation event – that is, reduce it (final sentence of clause X2.1). But note here we are referring only to the employer's contributions to national insurance, not to that of an employee.

180. Applying Option X2 (Option X2)

QUESTION

We are having a debate with our *Project Manager* on our ECC Option A contract. Option X2 was included but the *Project Manager* is saying that we should have predicted changes to legislation prior to the Contract Date and they should have been included in the tendered total of the Prices. Is the *Project Manager* correct?

ANSWER

No, the *Project Manager* is not correct. Clause X2.1 clearly states that a change in law is a compensation event if it occurs after the Contract Date. Whether you could predict this occurring or even categorically knew it would occur is not relevant here. If a change in law occurs after the Contract Date, then it is a clear compensation event. This could either increase or reduce the Prices.

181. Paying in different currencies (Option X3)

QUESTION

We are embarking on a processing plant in Yorkshire (UK) and we need to pay partly in Swiss francs, euros and pound sterling. I appreciate that Option X3 is most appropriate, but I am unsure how this works – can you explain how I complete?

ANSWER

Yes, it is quite straightforward. You need to define the *currency of the contract* – presumably this will be pound sterling. You will then need to define which items and activities will be paid in other currencies, what each currency is and the total maximum payment in that currency. It may be prudent to include an assessment for potential compensation events as this amount cannot be exceeded.

The final aspect to complete is the date when and by whom the indices are published. This will then allow the *Contractor* to assess this as part of its tender.

182. Ultimate holding company guarantee (Option X4)

QUESTION

We need to include an ultimate holding company guarantee on our ECC project but cannot find how to complete this in the Contract Data – can you advise on the wording of the ultimate holding company guarantee?

ANSWER

The NEC does not provide a sample ultimate holding company guarantee as this is something that is typically drafted on a bespoke basis by each *Client*. You will need to seek specialist advice on this. The wording of this should then be included in the Scope – this avoids any debate on the wording on award. The ultimate holding company guarantee is then either provided at the Contract Date or within 4 weeks of the Contract Date. Note that this documentation is important to have in place – this is reinforced by virtue of clause 91.2 (R12). This states that the *Client* may terminate the *Contractor's* obligation to Provide the Works if this is not provided within the 4 weeks. 'May' means that this is discretionary, so the *Client* has the option.

183. Is there a standard ultimate holding company guarantee form? (Option X4)

QUESTION

We are in discussion about a project using the Professional Service Contract (PSC), and there is a requirement for a parent company guarantee under Option X4. Is there a form provided by the NEC for this?

ANSWER

There are no standard forms for such a guarantee, as each organisation tends to use its own version. The version of the guarantee required by your potential *Client* should be set out in the Scope: see the wording of clause X4.1 – 'in the form set out in the Scope'. If there is no form in the Scope, you will need to agree the wording with the *Client*.

184. Defining *sectional* Completion (Option X5)

QUESTION

We need to have a *sectional* Completion on a project we are procuring under Option A of the ECC. The project involves the building of a new office block (for which we need sectional Completion), and the *Contractor* will undertake office refurbishment once we have vacated. We are incurring cost for every day we are renting the existing offices, so I both want to incentivise the *Contractor* to complete early and protect ourselves against the cost of delay if it is late.

My query is, how should we draft this in the contract to meet our objectives?

ANSWER

This is straightforward to incorporate into the ECC. Secondary Option X5 should be selected, and the description of the *section* carefully drafted. You need to describe Completion of the *section* and the Completion Date. You may wish to consider using Option X6 (bonus for early Completion) and Option X7 (delay damages) to provide the incentive of early Completion and protection against additional costs if the *Contractor* is late. Under UK law, the damages need to be a genuine pre-estimate of loss.

Finally, it is advisable to follow the Scope guidance in volume 2 of the ECC User Guides – this provides support in precisely establishing the definition of *sectional* Completion.

185. How is *sectional* Completion confirmed? (Option X5)

QUESTION

We are currently using an ECC contract on a project that contains *sectional* Completion, all as outlined within secondary Option X5 of Contract Data part one.

Could you confirm how and in what form *sectional* Completion is to be confirmed?

ANSWER

You deal with *sectional* Completion in the same way as Completion (clause X5.1).

Completion is a defined term (clause 11.2(2)), and that definition applies to Completion of any *section* of the *works* if Option X5 is used. The *Project Manager* decides the date when Completion (as defined in the contract) of any *section* occurs and certifies it within 1 week of that date (clause 30.2). Certificates must comply with clauses 13.1, 13.6 and 13.7.

186. *Sectional* Completion – effects of delay (Option X5)

QUESTION

We are a *Contractor* under an Option C contract that contains Options X5 and X7. The delay damages for the relevant section are £9000 per day. We planned to meet the *sectional* Completion Date some 7 days early but have been delayed by another *Contractor* (employed by the *Client*) not providing us with information that we stated on the Accepted Programme. The *Project Manager* is stating that this float is 'shared', and as such it can use it. Can you advise where we stand with this and the appropriate clauses?

ANSWER

This is the same situation as if you planned to complete the whole of the *works* early and then you were delayed – this would be a compensation event. Clause X5.1 confirms that each reference to Completion (unless stated as the whole of the *works*) applies equally for the whole of the *works* or any *section*. This is a compensation event under clause 60.1(5). Clause 63.5 then confirms how this is assessed. If you planned to achieve Completion earlier than the *sectional* Completion Date and you were delayed by Others, then by whatever amount this is added to the *sectional* Completion Date.

187. One *defects date* from Completion of the whole of the *works* (Option X5)

QUESTION

I attended an NEC training course, and I am sure that it was mentioned that you do not have different defects liability periods running from each *sectional* Completion. However, I cannot find where this is mentioned in the contract. Can you advise?

ANSWER

That is correct – the ECC does not have multiple *defects dates* running from each sectional Completion. Contract Data part one states a period between Completion of the whole of the *works* and the *defects date* (typically 52 weeks).

188. How to calculate the bonus (Option X6)

QUESTION

We are a major supermarket client and are about to use the ECC for the first time. I note the ability to insert a bonus for early Completion, which would fit with our objectives as the *Client*.

Is there a defined method of calculating the bonus for early Completion or can it just be a figure that we come up with?

ANSWER

The bonus for early Completion (Option X6) can be any figure you wish. Obviously, for internal probity/governance this will need to be less than the benefit you get from early Completion.

Related to this is the need to have a clear definition of Completion. The default is that the *Contractor* has corrected Defects that prevent the *Client* using the *works* and Others from doing their work. It is recommended that you follow volume 2 of the ECC User Guides on how to write the Scope. This allows you in section WI 405 to define precisely what you expect at Completion. This may include operational and maintenance manuals, refrigeration units being tested, as-built drawings being handed over and user training on the new facility, for example.

189. Applying delay damages (Option X7)

QUESTION

I am a *Project Manager* on an ECC Option B contract. The *Contractor* is 20 days late on Site and we have delay damages of £3000 per day. The *Contractor* has been extremely effective on the project, and we do not want to take the damages from it. Am I able, as the *Project Manager*, not to deduct the damages?

ANSWER

As a *Project Manager* under the ECC you are obliged to act as stated in the contract (clause 10.1). This covers everything, and, unfortunately, you cannot select which clauses to follow and not to follow. If delay damages are included and the *Contractor* achieves Completion after the Completion Date, then delays damages are paid by the *Contractor* at the amount stated in the Contract Data multiplied by the number of days.

The only thing you can do is recommend to the *Client* that damages are not taken, but any certificate that you issue must account for the deduction of the delay damages. The *Client* may then choose how much to pay the *Contractor* in addition to the certified sum. Under clause 12.3, if the Parties (i.e. the *Client* and *Contractor*) agree, then it can be put in writing and signed that delay damages are not deducted – but this is not something that, as the *Project Manager*, you can decide yourself.

190. Dealing with late Completion (Option X7)

QUESTION

Under an ECC Option A contract, how would the *Project Manager* deal with a *Contractor* finishing late when no compensation events are apparent? Under the Joint Contracts Tribunal (JCT) contracts, the Contract Administrator would issue a Certificate of Non-Completion. Is this the case with the ECC also?

ANSWER

The ECC mandates the Parties to notify and deal with compensation events as they are known about. The processes and timescales are set out in clauses 61 to 66. And, as you seem to realise, the primary way that the Completion Date can be extended to a later date is through that compensation event procedure.

We will assume that you have included the delay damages Option X7 in your contract. There is no such thing as a Certificate of Non-Completion in the ECC, because none is needed. That is because, unlike the JCT forms of contract, it is the *Project Manager* that deducts delay damages as part of its assessment of the amount due to the *Contractor* (clauses X7.1, 50.1, the third bullet of clause 50.3 and clause 51.1). So, all that is needed is that, in the certificates issued after the Completion Date, the *Project Manager* deducts the delay damages accordingly.

We would also recommend that the *Project Manager* notifies an early warning, calls an early warning meeting (if this has not already happened) and tries to set about solving the problem of delay, if possible. Completion is a defined term (clause 11.2(2)). The *Project Manager* decides when Completion is achieved and certifies that date (clause 30.2). In addition, the *Project Manager* must decide when take over has occurred and certify that as well (clause 35).

191. Delay damages and take over (Option X7)

QUESTION

We are a *Contractor* on an Option D project. The delay damages are £90 000 per day. We were 3 days in delay when the *Client* started to take over and began using part of the *works*. How does this affect the delay damages and how should this process be recorded and managed?

ANSWER

For the first 3 days the full £90 000 per day will clearly apply (clause X7.1). The *Client* is entitled to take over any part of the *works* before Completion has been certified. The *Project Manager* certifies this within 1 week of the date. If the *Client* takes over before both Completion and the Completion Date, then this would be a compensation event. This is not the case on your project, as you have gone past the Completion Date.

The bad news is that the *Client* taking over is not a compensation event. However, the good news (if we may call it that) is that the delay damages are reduced by the proportional benefit obtained by the *Client* in taking over. This is assessed by the *Project Manager*, and typically may represent a significant reduction in the damages, as most clients take over those parts of the *works* that offer most benefit.

192. Delay damages and gain share (Option X7)

QUESTION

We have an ECC Option C contract. The *Contractor* has achieved Completion after the Completion Date. The delay damages are £300 per day but the *Contractor* has also achieved a gain share. How do we calculate this – do we deduct damages from the cost first, then calculate the gain share?

ANSWER

No, the calculation is based on the difference between the Prices and the Price for Work Done to Date (clause 54.2). The Price for Work Done to Date (clause 11.2(31)) under Option C is based on Defined Cost (clause 11.2(24)). The amount due (clause 50.3) is a mixture of the Price for Work Done to Date, plus amounts to be paid to the *Contractor* (i.e. the share) less amounts to be paid by the *Contractor* (i.e. the delay damages). So, to answer your question, work out the share first then deduct the damages after.

193. Undertakings to the *Client* or Others (Option X8)

QUESTION

As a *Client,* we often need to amend contracts to address collateral warranties. How is this dealt with in the ECC?

ANSWER

This is dealt with under secondary Option X8 and referred to as 'Undertakings to the *Client* or Others'. Others are defined under clause 11.2(12).

The undertakings are stated in the Contract Data and the wording is included in the Scope. Under clause X8.4 the *Client* prepares the undertakings and sends them to the *Contractor* for signature. The *Contractor* then returns them to the *Client* within 3 weeks (clause X8.4).

As this is a specialist field, we advise that legal advice should be sought when drafting the undertakings.

194. Applying collaboration principles (Option X12)

QUESTION

We want to incentivise our *Contractor* to surpass the basic contract requirements on our ECC Option C project. We are also a strong advocate of collaborative working. I presume that Option X12 is the appropriate Option to use – and can you provide guidance on how this should be completed?

ANSWER

Option X12 does not seem like the most appropriate Option for you. It is for arrangements in which multiple partners are involved – such as on complex projects when the *Client* wants perhaps its consultants, the main *Contractor* and key members of the supply chain to work together closely. An *incentive schedule* is completed, and a Core Group is established to focus effort.

In circumstances when you just want to provide incentive to the *Contractor* (i.e. no other partner is involved), then Option X20 is the most appropriate. You will need to append an *incentive schedule* – this needs to clearly define Key Performance Indicators, how these will be measured and what the incentive mechanism is.

195. Getting performance bonds in place (Option X13)

QUESTION

As a local authority, our standard tendering documents require that we include a performance bond. Where should the wording for this be included and how do we ensure that the bond is provided after contract award, as we tend to find that contractors are reluctant to provide this once the tender has been won?

ANSWER

It is recommended that you follow volume 2 of the ECC User Guides on how to draft the Scope. Specialist legal advice will need to be sought as to the wording itself.

In terms of making sure that this is signed, you could get this signed on contract award. Failing that, there is a strong provision in the contract – clause 91.2 (R12) states that if the *Contractor* does not provide a bond that the contract requires, the *Client* may terminate the *Contractor's* obligation to Provide the Works. In accordance with clause X13.1, 'may' implies discretion, so it is up to the *Client* whether it exercises this right.

196. Paying in advance (Option X14)

QUESTION

We are using Option X14 (advanced payment to the *Contractor*) on a nuclear project in the UK. We recognise that the *Contractor* must make significant investment up front in order to meet the project timescales. Our question is, does the amount include VAT and what paperwork needs to be in place before the advanced payment is made?

ANSWER

You need initially to define the advanced amount – it should be made clear whether this includes VAT. If due on amounts to be certified later, then it would also be due on advanced payments. You will also need to define the instalments for repayment by the *Contractor*.

Finally, you will need to establish whether an advanced payment bond is required or not (clause X14.1). The bond should normally be provided before the Contract Date. The advanced payment can be delayed until not later than 4 weeks after the *Client* has received the bond. But where a bond is required, the advanced payment will not be made by the *Client* until the bond has been delivered.

197. Delayed advanced payment! (Option X14)

QUESTION

We are a *Contractor* on an Option A ECC project. We have just started on a high-value shopping centre complex. The contract includes an advanced payment but, due to administration problems within the *Client's* organisation, the advanced payment is late in being made by a month. Where do we stand with this – are we entitled to interest, and what about the disruption this has caused us?

ANSWER

Clause X14.2 states that any delay to the advanced payment is a compensation event, and as such would be subject to the same assessment rules as any other compensation event – that is, the cost and time both potentially need to be assessed.

198. Design liability (Option X15)

QUESTION

I am completing Contract Data part one on our ECC Option B project. This is a civil engineering project with some tunnel design required by the *Contractor*. I understand that the inclusion of Option X15 is appropriate. What does the *period for retention* mean, and how long should this be?

ANSWER

Clause X15.4 states what the *Contractor* should retain for the *period for retention*. This could include things such as data, drawings and reports. It is up to the *Client* how long these items should be kept for – there is no hard and fast rule here.

199. Reasonable skill and care (Option X15)

QUESTION

We are a *Contractor* delivering a complex prototype project in the nuclear sector (ECC Option C). Part of the *works* is unique and, to the best of the knowledge of both us and the *Project Manager*, has never been undertaken before – and Option X15 applies. In one aspect of the design, both ourselves and the *Project Manager* agree that reasonable skill and care has been exercised by us in the design, yet a Defect has still occurred. Our question relates to how this is dealt with – is this a compensation event?

ANSWER

Under Option X15.1, unless you failed to carry out the design using the skill and care normally used by professionals designing works like the *works*, the *Contractor* would not be liable for a Defect.

200. Subcontractor retention (Option X16)

QUESTION

The *Project Manager* that we are dealing with on an ECC option C contract is not including the amounts of retention we have deducted from payments to Subcontractors in its assessments of the Price of Work Done to Date, as it says we are confusing the term 'retention' when phrased in the main contract clause 11.2(24) as having some connection with 'retention' in the subcontract. The *Project Manager* believes that these are entirely different concepts resulting from entirely different contracts. The *Project Manager* states that Defined Cost at a main contract level should not have main contract retention deducted, as main contract retention is deducted from the Price of Work Done to Date if Option X16 is used. However, the *Project Manager* has also stated that any subcontract retentions should not be included in Defined Cost, as these are an entirely different concept from the main contract retention and says that we should only be paid the 'amount of payments due to Subcontractors'. What do you think?

ANSWER

We believe the *Project Manager* is incorrect; all that you need to do is to read and understand clause 11.2(24). The first part of the first bullet makes it clear that 'payments due to Subcontractors for work which is subcontracted' is being referred to here, and the deductions listed subsequently in the sub-bullet points can only refer to deductions made from those payments.

If you look at the rest of the list, many of them can only apply to the payments you are making to the Subcontractor, as it is clearly passing on things that you have had deducted – for example, the second sub-bullet, or you have supplied to the Subcontractor – for example, the last sub-bullet.

The only way that the *Client* can deduct retention from you is if it uses Option X16. It cannot introduce it through the back door in this way. The ECC User Guides make it clear that this refers to deductions from the Subcontractor's payment, not the *Contractor's*. It says that they are 'normally' deducted, but, in the case of the first one, that will only be the case if the *Client* uses Option X16. Just because it has not done what may be 'normal' does not change the meaning of the clause.

201. Retention percentage (Option X16)

QUESTION

We are a *Contractor* on an Option B ECC project. Contract Data part one states on the initial page that Option X16 will apply; however, the *retention percentage* was not completed by the *Client* at the tender stage. The *Project Manager* is now stating that normally this is 5% or 3% and, giving us the benefit of the doubt, it should be 3%. Where do we stand on this?

ANSWER

There is no 'normal' percentage as such. The guidance in the Contract Data warns: 'Completion of the data in full, according to the Options chosen, is essential to create a complete contract.' If not completed in full you simply have an incomplete contract.

You cannot just fill in the rest of the Contract Data with supposed norms or industry practice. If legal advice was sought, then perhaps it would be found most favourable to yourselves as the *Contractor*, as we have one aspect of the contract mentioning retention and another section not. Perhaps the best way forward is to agree some sensible changes to the Contract Data to correct the drafting error – this can only be undertaken by the Parties (i.e. the *Client* and *Contractor*). It also needs to be in writing and signed by the Parties (clause 12.3).

202. Exceeding the retention amount (Option X16)

QUESTION

We have an ECC Option A contract – this is for the refurbishment of an existing swimming pool in a secondary school. For a variety of reasons, the project did not go well. The re-lining of the pool was ultimately unsuccessful. Completion was certified by the *Project Manager* some 3 months ago, and the *defects date* is 52 weeks after Completion. The original *Contractor* is refusing to come back to rectify notified Defects. Our problem now is that the likely costs of correcting the Defects exceeds the 1.5% retention that is currently being held. How do we get this money back from the original *Contractor*?

ANSWER

Firstly, it is good that all the correct procedures appear to be being followed – issue of a Completion certificate and Defects being notified. It is important to maintain the notification of Defects (as per clause 43.2) even if the *Contractor* is not coming back. The contract states that (under clause 46.1) if the *Contractor* does not rectify the Defect within the *defect correction period*, then the *Project Manager* assesses the cost to the *Client* of having the Defect corrected by other people. The *Contractor* pays this amount. The contract does not say that you have to have the Defect corrected, so you could just retain the money – presumably for a leaking pool the remedial work needs to be undertaken!

In terms of getting your money back, the *Project Manager* will need to assess the amount due under clause 50.4 at the next assessment date. This is a mixture of Price for Work Done to Date plus other amounts to be paid to the *Contractor* less amounts to be paid by or retained from the *Contractor*. If the balance of monies is now such that the *Contractor* owes you money, then that is what the *Project Manager's* certificate will reflect. It is worth noting that clause 51.2 is written in a 'neutral' way in that failure of the *Contractor* to pay an amount it owes you would accrue interest. You will need to get your finance department to raise an invoice – if the *Contractor* fails to pay this, then adjudication would be the next step before the normal debt recovery procedures could be followed.

203. Deciding the *retention free amount* (Option X16)

QUESTION

In using the ECC with Option X16 retention, can you advise on how the *retention free amount* is calculated?

ANSWER

The term *retention free amount* is in italics, which means that it is an identified term (clause 11.1), and the amount is therefore identified in the Contract Data. So, the amount will be decided by the *Client* at the tender stage and inserted in Contract Data part one, under the statements for Option X16. It is entirely up to the *Client* how much this amount is and how it is calculated.

204. Release of retention (Option X16)

QUESTION

We have an ECC contract incorporating Options X5 (sectional Completion) and X16 (retention). There are four *sections* defined, and the retention percentage is 3%, with nil *retention free amount*. Three of the four *sections* are complete and in use by the *Client*. At present, one-half of the retention has not been released in respect of the three completed *sections*. Retention is therefore currently held at 3% on the full value. The *Contractor* is of the view that one-half of the retention applicable to the three *sections* completed should be released. The reason that the retention has not been released is that clause X16.2 refers to the amount retained being halved at Completion of the whole of the *works*.

The whole of the *works* will not be complete until the final phase is complete, so, until then, the retention remains at 3%. Clause X5.1 by reference to 'unless stated as the whole of the *works*' apparently excludes the application of the retained sum from each *section*. We think the *Contractor's* interpretation perhaps reflects provisions of other standard forms of contract, but what is your view?

ANSWER

It is completely irrelevant what any other contract says (or not) about retention – it is all about what your particular contract says, and we cannot fault your interpretation. If you simply read the relevant provisions through with the *Contractor*, the logical conclusion is that retention is not released for each *section*, only at Completion of the whole of the *works*.

205. How to draft low-performance damages (Option X17)

QUESTION

We are considering using Option X17 on an Option C ECC project. This is for the construction of a new energy centre on our healthcare estate.

Can you advise how this should be drafted?

ANSWER

The purpose of low-performance damages is to reimburse the *Client* where the performance of the *works* in use fails to reach a specified level due to a design or other fault by the *Contractor*. The required output should be clearly specified in the Scope. It is advisable to follow the structure proposed for the Scope in volume 2 of the ECC User Guides.

You will also need to complete the low-performance damages section in the Contract Data.

The low-performance damages for, say, an energy centre could be graded based on certain levels of low performance – for example, a failure to operate at the capacity stated or for the fuel input specified. In each case, the *Client's* future costs could be computed into a present value.

The deductions for low-performance damages are made in the Defects Certificate.

206. Limiting liability (Option X18)

QUESTION

Are delay damages included within the limitation of liability Option? That is, are delay damages capped at a certain level?

ANSWER

No, under clause X18.5 there are a number of 'excluded matters' (i.e. that are not subject to a limit of liability); these include (among others) delay damages and low-performance damages.

207. *End of liability date* (Option X18)

QUESTION

We are a *Contractor* bidding for work that includes Option X18. We are trying to understand the significance of the *end of liability date*. This is completed at 10 years after Completion of the whole of the *works*. Is this the end of all our liabilities?

ANSWER

Yes, after this point the *Contractor* is not liable in contract to the *Client* for any matter, which would include a Defect that is notified after this date.

208. Providing incentive (Option X20)

QUESTION

We have an ECC Option A project. We are a pharmaceutical client with tough targets to achieve on our project. We want to provide incentive if the *Contractor* betters the targets and provide penalty if it fails to achieve them. Option X20 refers to an *incentive schedule* – presumably this is attached and covers both incentive and disincentive?

ANSWER

No, Option X20 is purely there to incentivise. The *incentive schedule* should be appended to Contract Data part one but should contain only incentives; these could incentivise the *Contractor* with, say, additional work on a framework or monetary payments. The Key Performance Indicators should be clearly defined, as should the method of calculating.

If you want to provide a 'penalty'-type provision, then the correct secondary Option is Option X17 – low-performance damages. Under UK law, these should be a genuine pre-estimate of loss as penalties for breach of contract are not enforceable.

209. Using Option X20 (Option X20)

QUESTION

Do the Key Performance Indicators in Option X20 have to provide incentive, or can they be used purely for monitoring purposes with potential for future work?

ANSWER

The intent of Option X20 is to provide the *Contractor* with an incentive. Some *Clients* just use Option X20 purely for monitoring purposes. The Key Performance Indicators are often aligned with corporate objectives. The Key Performance Indicators do not necessarily have to provide direct financial reward; they could offer potential for, say, future work with the *Client*. We suppose we must respect the fact that the less direct a reward is, the less of a true incentive it provides.

210. How to build in potential extensions to the Service Period under the Term Service Contract (TSC) (Option X23)

QUESTION

We are about to tender for services using the TSC. The initial Service Period is for 5 years with the potential to extend for two 2-year periods based on certain performance criteria being achieved. How do we achieve this in the TSC?

ANSWER

This is provided for in secondary Option X23 in the TSC, so a *Client* may use this where appropriate.

Clause X23.1 allows the *Client*, with the agreement of the *Contractor*, to extend the Service Period. The *Client* notifies the *Contractor* and the *Service Manager* before a *notice date*. Sensibly, the *notice date* provides advanced warning – this could be, say, 3 or 6 months. Each *period for extension* and *notice date* is completed when Option X23 is filled out. So, in your case, there would be two *periods for extension* – each of 2 years.

Clause X23.2 allows *criteria for extension* to be identified in the Contract Data. These could be, for example, the attainment of minimum key performance indicator thresholds.

Similar provisions apply in the NEC4 Facilities Management Contract.

211. How to 'close out' each financial year under Option C of the TSC (Option X24)

QUESTION

We frequently use main Option C of the TSC but need each financial year to be dealt with separately. As an organisation, we need finality each financial year and effectively 'close out' each year's accounts.

How is this dealt with under the TSC?

ANSWER

This provision is dealt with as a secondary Option under the TSC, so a *Client* may use this where appropriate. The *accounting period* may be any period stated, and is completed in the Contract Data (i.e. April–March or January–December).

Option X24 (*accounting periods*) does exactly what you are after. Clauses X24.1 and X24.2 require that the *Service Manager* assesses the 'final' amount due within 13 weeks of each *accounting period*. This is 'conclusive evidence' (clause X24.2) of the final amount due unless this is disputed.

Clause X24.3 covers situations where an item in the Price List straddles two *accounting periods*. In this case, a proportion is assessed that has been completed in each respective *accounting period*.

Finally, it is worth noting that Option X24 may be used under any of the main Options A, C or E.

Similar provisions are included in the NEC4 Facilities Management Contract.

212. Is a 'day' a day? (Option Y(UK)2)

QUESTION

We have recently started using the PSC and would like to know if 'days' used in time periods in the PSC means calendar days or working days?

ANSWER

The PSC only uses days, as opposed to weeks, to describe time periods in two places: the dispute resolution Option W2 and secondary Option Y(UK)2. Both are only used for work subject to specific UK legislation, as set out in the preamble at the beginning of both Options, and to comply with that legislation.

In both cases, clauses within the Options clearly describe exactly what is meant by 'days' in the Option (clause W2.2(2) for Option W2, and clause Y2.1 for Option Y(UK)2). Although the wording is slightly different, the meaning is the same.

The only other time you may come across the word 'days' is if the *Client* has used it when describing time periods in the Contract Data or Scope. The meaning will depend on exactly what the *Client* has written. If the *Client* has just used the word 'days' without any other descriptor or qualifier, then it will mean exactly what is says: days. Saturday and Sunday are days, as are public holidays, so they all count.

213. The purpose of Option Y(UK)3 (Option Y(UK)3)

QUESTION

Can you explain the purpose of Option Y(UK)3?

ANSWER

Option Y(UK)3 is only applicable on projects in England, Wales or Northern Ireland. This secondary Option ensures that only the terms set out in Contract Data part one can be enforced by those persons, or class of persons, that are named, called a *beneficiary*. It basically ensures that third-party rights cannot be implied from any other terms of the contract or Scope.

With minor amendments Y(UK)3 may also be used in Scotland, although the relevant legislation differs from that in the remainder of the UK.

214. Preparing quotations for compensation events (clause 11.2(23))

QUESTION

We are using an ECC Option A contract. The contract is silent on whether the *Contractor* can recover the cost of preparing quotations for compensation events. Would time spent obtaining quotations for compensation events from Subcontractors be allowed to be included in a quotation?

ANSWER

Yes, obtaining quotations from Subcontractors is part of preparing a quotation, and if you can demonstrate that this leads to an effect on Defined Cost under clause 63.1, then it can be included in the quotation.

215. Increase of resources by the *Contractor* (Option C)

QUESTION

In an ECC Option C contract, the *Contractor* is in 'pain' and will not recover the situation. The *Contractor* seems to be undertaking every administration activity possible on Site in what appears to be a move to make money on staff who would not normally operate on Site. Is it contractually acceptable for a *Contractor* to relocate staff that would normally be head office based (i.e. accounts staff) to the Site to recover the cost of such people through the SCC?

There also seems to be an increase in the attendance of the Site-based quantity surveying staff from one to three days in order to do the same. There does not seem to be a proportional increase in performance or output from the surveyor related to our project, and we believe this is an attempt to increase earnings.

ANSWER

The *Contractor* is entitled to organise its staff as it sees fit. And it is entitled to be paid for its people when they are in the Working Areas if they are Providing the Works, which, because of the definition in clause 11.2(15), would apply to these people. Of course, the more the *Contractor* increases the Price for Work Done to Date, the more it will end up paying and the greater its 'pain'. Therefore, this is a shared risk, and, in practice, the *Client* will receive a proportion of these Defined Costs at the end of the project.

216. Disallowed Cost (Option C)

QUESTION

We have an ECC Option C contract and are in a situation whereby a large metal sliding door has been delivered to the Site in a damaged condition. Putting to one side the failings of the supply chain and all the quality control procedures, the *Contractor* then proceeded to install the door. The *Supervisor*, as duly obliged, notified a Defect on becoming aware of the damage (i.e. after installation). Under the terms of the contract, the cost of rectification of Defects before Completion sits with the *Client*. Given that the door was delivered in a damaged condition, does this still apply in this instance? If yes, then if the door had been rejected at the Site and returned straight back to the supplier for repairs or provision of a new door, would the *Client* still be liable for the costs of this? Note that there is currently no direct sight of the supplier's terms and conditions but is this necessary given the condition of the goods supplied and that the door still performs to its functional requirements?

ANSWER

We are struggling to understand your suggestion that this problem sits with the *Client*. Option C is, in fact, a shared risk contract, and the risk of Defects before Completion is one of those shared risks. This clearly seems to be a Defect, and the *Contractor* is obliged to correct that Defect. The question is, who pays what and when? Normally, the *Client* pays the cost, and therefore that increases the Price for Work Done to Date. However, the target stays the same (because this is not a compensation event), so the *Client* recovers some of that back through the pain–gain mechanism on Completion. However, in this case, if the door was obviously damaged, then the *Contractor* should have given the *Client* an early warning of that under the fourth bullet of clause 15.1. Since it has not, then one of the definitions of Disallowed Cost (clause 11.2(26)) is any cost incurred only because the *Contractor* did not give an early warning that it should have. So, the additional cost of dealing with this problem now rather than before the door was fitted is a Disallowed Cost. Thus, in theory, at least the cost of returning the door to the manufacturer before it was installed to get it sorted out would be paid. However, that is going to depend on the terms of the contract that the *Contractor* has with its supplier, so you will need to see those. If the door was Defective on delivery, then that will normally be down to the supplier to sort out. The *Contractor* must keep those records and show them to the *Project Manager* under clause 52.

217. Productivity of resources (Option C)

QUESTION

Our contract is an ECC Option C project, and our question relates to the SCC Disallowed Cost. Item 1 of the SCC deals with people in the Working Areas. We have had it put to us by the *Project Manager* that our personnel have been (despite not actually providing the names of any individuals as evidence) taking too long for breaks; walking to other areas of the Site, to cabins and/or to clocking stations is taking too long; they have been standing around; and they have left the workface early and been waiting at the clocking stations to clock out for periods of time. The *Project Manager* has stated that, in reference to item 11 of the SCC, 'time not worked' is therefore not payable. Our interpretation of matters is that this is part of productivity or potential lack of it, which is catered for in the *Contractor's* share. It is also our opinion that the interpretation of the above is that it gives a greater emphasis to the 'amounts paid by the *Contractor*' (i.e. we get paid for the amounts we have paid).

A secondary argument put to us is that as the activities are 'not used to Provide the Works', they are not part of Defined Cost. Again, our interpretation is that this is in fact a resource used to Provide the Works – as perceived lack of productivity is still Providing the Works, just not as effectively as we would wish.

Your guidance would be appreciated on the above, as it would appear that the *Project Manager* is attempting to remove substantial costs, which have been paid from the Price of Work Done to Date, to ensure that this does not even reach the calculation for the *Contractor's* share and that, therefore, the *Client* benefits in full from these deductions rather than sharing in the construction/productivity risk we believe this form of contract caters for.

ANSWER

Our view is that the *Project Manager* is unable to do this. To deal with the first argument – that is, 'time not worked' – the first thing to say is that this is misquoted. Item 1 of the SCC in fact refers to 'the cost of people … proportionate to the time they spend working in the Working Areas'. This merely conditions who this refers to, and undoubtedly the people involved fall within that category. It does not say that if they are paid despite not working it will not be allowed. And that is clear once you see the other matters that are paid for, such as travel time, absences for sickness and holidays, bonuses, incentives and so on, which are all payments made that do not relate directly to the time worked. In fact, some of them, such as travel, holidays and sickness, are specifically about 'time not worked' – and yet they are still paid. In addition, you could successfully argue that even if they were not working, then you are paying them for not working as a 'special allowance'. Alternatively, you could successfully argue that the time taken to get to the canteen or wherever was a payment made in relation to travel, as it does not specify what travel or to where. The simple fact is that if you paid these people in accordance with their terms and conditions, and normal construction industry practice, you are entitled to be paid for them as Defined Cost, and the *Project Manager* is, frankly, wrong to suggest otherwise.

Regarding the second point, again that seems incorrect. These people were used to Provide the Works and have been paid in accordance with their terms and conditions. And the costs therefore were incurred to Provide the Works. The only way that the *Project Manager* could claim this would be to quote the penultimate bullet point of clause 11.2(26), concerning Disallowed Cost. However, even that would be a bad point because it is quite clear that you must allow for 'reasonable availability and utilisation'. And this simply gets nowhere near jumping that hurdle. The *Project Manager* has clearly misunderstood how the Option C aspect works. As you correctly say, productivity issues such as these are a shared risk if productivity is low and shared reward if it is high. That is the whole point of using Option C.

218. Recovering possible accrued cost in an application for payment (Option C)

QUESTION

I would like some clarification on the inclusion of forecast of accrued costs in an application for payment. By this, I mean costs that probably will be paid for in the following month, for which the cost is not exactly known and has not been paid yet – for example, re-fuelling company cars for people who are 100% allocated to the project. In theory, the *Contractor* has an idea through historical records of how much per month this generally costs, but this cannot be substantiated yet. If we also read about a likely fuel cost increase in the month, can that be allowed for too?

I believe neither cost should be included, because, at the point in time of the application, the costs cannot be substantiated. Is this correct?

ANSWER

We consider that your opinion is correct. But the *Project Manager* should look at how stable and predictable previous payments have been. For example, salary payments to people committed to the project full time or hire charges for Equipment such as site offices and welfare facilities will be predictable and previous months' audits will have identified suitable supporting accounts and records. But items such as fuel card expenditure or speculative agency people provision will need to await accounts and records to support their inclusion as Defined Cost.

With the second category of costs, these will work their way through in time as being valid Defined Cost, assuming it is at market rates, but right now you have no accounts and records for this. One of the definitions of Disallowed Cost is any cost that cannot be justified by the *Contractor's* accounts and records. These sorts of costs cannot be justified by those records at the point when the *Project Manager* is making its assessment, so are disallowed. Of course, for the next assessment, when the accounts show such payment, this will be allowed, even if they have not yet been paid (if they will be paid by the next assessment of course). This provision covers accruals made in the *Contractor's* accounts for items that it has invoiced for (or other justification such as a Subcontractor payment certificate) but that have not yet been paid.

219. The use of the Activity Schedule (Option A)

QUESTION

The tender documents were sent out and included an Activity Schedule and a suite of drawings. The successful tenderer included a Bill of Quantities in its tender reply, summed to a summary sheet based on the Activity Schedule. For some reason, the *Contractor* priced some bill items that do not appear on the drawings. The *Contractor* is now asking for payment of the lump sum for the activity, which includes sums for items of work that have not been undertaken. We are acting as the *Project Manager* and consider that if the *Contractor* has not done the work, it is not due the money. The *Contractor* argues that including the Bill of Quantities in the tender return was a mistake. I consider the bill to be a proxy for the *Contractor's* duty to split the activities (groups of activities) into smaller packages to avoid conflicts of this very sort.

ANSWER

Option A is based on an Activity Schedule, and that alone. It has no Bill of Quantities in it, and there is no use for such an item anywhere. It should not have been included in the contract because it creates ambiguity, although the status of either the Bill of Quantities or the Activity Schedule is not at all clear here. If the Bill of Quantities was not included in the contract, then it is, in any event, irrelevant.

Having said all of that, the contract is clear – it is the activities and prices in the Activity Schedule that matter, and you cannot look to the Bill of Quantities. It is the Activity Schedule and only that, which states what the *Contractor* is to be paid once it carries out the works in the Scope. How and what it has allowed to get to that figure is irrelevant. Whether it has allowed too much or too little for whatever reason is its good or bad fortune. In order to demonstrate that, we would ask you to consider what would have happened if the *Contractor* had forgotten to include in its Bill of Quantities (and therefore its Activity Schedule) work that was in the Scope? If it asked you to pay for them as an extra, we suspect that you would say no, which would be the correct answer. It is the same the other way round.

220. Items on an Activity Schedule not undertaken (Option A)

QUESTION

We are working on an ECC Option A contract. During the tender process and after inspection of the Site, the *Contractor* allowed for an element of hard breakout that was perceived as required in order to carry out the *works*. This amount was acceptable and included within the Prices with an annotation to describe the *works* as a description within a pricing document that was subsequently included in the Scope. During the *works*, the hard breakout was not required, as the assumed slab formation did not exist. Is the *Project Manager* correct in assuming these *works* can be omitted from the contract? And if so, could you advise how this would be measured.

ANSWER

Option A is usually quite clear – it provides for a lump sum price to cover all the work in the Scope. The main way that the Prices will change is if there is a compensation event. It is not subject to remeasurement or any 'provisional'-type items. This clarity seems to have disappeared in the way you have wrongly set the contract up.

Firstly, how you describe things in the Activity Schedule is irrelevant to what you are required to construct (see clause 20.1 and then clause 55.1). And, secondly, the Scope does not state what you get paid – only the Activity Schedule does that (clauses 50.3 and 11.2(29)). Further, the words in the Scope only form the Scope when they comply with the definition in clause 11.2(16). Somehow you have managed to muddle the two different documents together, and it is difficult to suggest how this might be interpreted.

The answer may depend on what the words you put into the Scope say. They could be interpreted as meaning that if you do the hard breakout, then you get paid, but if not, you do not get paid – which is in direct contradiction to the principles of Option A. They could equally be interpreted as being irrelevant because they do not comply with clause 11.2(16), so the amount of work would remain your risk (or reward). There is no clear-cut answer here, unfortunately, and the answer may depend on a legal analysis of the position.

Under clause 55.5, the *Contractor* can correct the Activity Schedule so that the activities on the Activity Schedule relate to the Scope. This cannot lead to a change in the Prices, though. It is not clear if the Activity Schedule does or does not relate to the Scope in this instance.

Finally, it might be worth asking yourself what you meant when you put the words in there. And before you answer that, ask yourself whether the answer would be the same if you had found twice as much hard breakout as was allowed for. The same point should be raised if it was the *Project Manager* asking the question. To hazard a guess, you would probably each be using the other's arguments! If you had wanted to deal with the risk of this concrete fairly and openly at the tender stage, then you should have not allowed for it in the Prices, and instead added the provision of a compensation event for finding it. It would then have been clear whether it was found would be at the *Client's* risk.

221. Additional management staff (Option C)

QUESTION

We have two questions listed below that we would appreciate an opinion on. We are using an ECC Option C contract.

1. The specification included in the Scope identified a particular type of hinge but the *Contractor* qualified its offer at tender for this element, which was not identified or agreed. The *Client* requires the originally specified hinge to be installed, and the

Contractor has now installed the hinge but has asked for a compensation event. This has been declined on the basis that the Scope has not changed – is this correct?

2. The *Contractor*, without an early warning or agreement, brought additional management staff in addition to the agreed management team. The application for its cost was declined as a Disallowed Cost on the basis that the resource was not agreed or necessary – is this correct?

ANSWER

The answer to the first question may well depend on how and where the *Contractor* 'qualified' its tender. We say 'may', because the answer may end up with a legal (as opposed to contractual) analysis, which we cannot provide.

Regarding the second question, the *Client* has no right to 'pre-approve' resources that the *Contractor* decides to bring to the Site. Option C is a risk-sharing contract. One of the risks you share is the *Contractor's* incompetence; conversely, one of the rewards you share is the *Contractor's* competence. It is for the *Contractor* to decide how best to manage the *works*, not the *Project Manager*. If these management staff were truly sitting doing nothing, then they could be a Disallowed Cost (penultimate bullet of clause 11.2(26)), but we doubt very much that you could show that to be the case. The *Project Manager* could request their removal and, a reasonable period after that, they would be a Disallowed Cost (see the wording of the same bullet). However, the *Contractor* would be right, then, to complain that the *Project Manager* is trying to manage the *works* on its behalf, and that could well turn into a dispute, so it is a dangerous path to tread. At the end of the day, the cost of these staff, like any benefits they bring, will be shared using the share mechanism at the end of the contract.

222. Revising the Activity Schedule (Option A)

QUESTION

A *Contractor* submits its tender for an ECC Option A contract. The *Contractor* is not familiar with the payment mechanisms and only allows for a handful of lump sum activities. Within the 'Preliminaries' section there is only one activity; the procurement and installation of sheet piling over the course of 2–3 months is another. These two activities amount to approximately one-third of the Prices.

The *Contractor's* tender is accepted, and at the start of the project it is advised that it will only be paid on completion of each activity. Realising that this may create problems with cash flow, the *Contractor* submits a revised Activity Schedule, with the activities broken down over the course of the programme. The *Client* is concerned about the closure of one of the *Contractor's* regional offices, although this is a major *Contractor*, and a parent company guarantee is in place. The *Client* does not believe that it is under any obligation to allow the Activity Schedule to be revised, and therefore rejects the *Contractor's* proposal.

Is the *Client* in breach of clause 10.2, in that it is not acting in the spirit of mutual trust and co-operation?

ANSWER

Contractually, it is not the *Client* that makes such decisions, it is the *Project Manager*. The contract does not allow for the changing of the Activity Schedule, unless it is changed in accordance with the contract (clause 11.2(21)). The only way it can be changed in Option A is set out in clauses 55.3 and 63.14, neither of which covers these circumstances.

The spirit of mutual trust and co-operation cannot be used to change a contract that the Parties have agreed to. So, there is no contractual obligation to change this Activity Schedule.

Having given you the contractual answer, our view is that the sensible and practical answer is somewhat different. In our opinion, it may not be in the best long-term interests of the *Client* for the *Contractor* to have a poor cash flow, with all the possible problems that will cause to the *Contractor* and the supply chain during the lifetime of the project. Our view is that it may be more sensible for the *Client* to enter discussions with the *Contractor* to see if agreement can be reached on a more sensible Activity Schedule, and then agree to change the contract in accordance with clause 12.3, but of course the Parties need to agree this, and it must not fall foul of any procurement laws.

223. Interest paid to a Subcontractor (Option C)

QUESTION

ECC Option C allows for interest to be paid on payments that are late. We are aware that the *Contractor* has made late payments to the Subcontractor, and subsequently the Subcontractor has been paid interest. Does this interest payment then become due between the *Contractor* and the *Client*?

ANSWER

The answer is going to depend on the terms of the subcontract, not the (main) contract. That is because you are required to pay what the *Contractor* has actually paid its Subcontractor (clauses 11.2(24), 11.2(31) and SCC), but only to the extent that it is required by the terms of the subcontract (second bullet of clause 11.2(26)). If the subcontract conditions require such interest to be paid, and if the *Contractor* has actually paid that interest to the Subcontractor, then the answer is yes. However, these are two very big 'ifs' – especially the second one. You should check to make sure that this amount has been paid, before paying it.

224. Open-book problems (Option C)

QUESTION

As a *Contractor*, we frequently work under the open-book main Options of the ECC. We often find that audits are undertaken late in the project and large sums of money are deducted by the *Project Manager*. This causes significant problems in terms of our company accounts and commercial reporting.

Our view is that regular audits should pick up any disparity. What does the ECC provide for here?

ANSWER

This is a historically recognised problem but should not be if the *Project Manager* undertakes regular audits of Defined Cost. The powers of audit of Defined Cost are granted in clause 52.3, and records to be kept stated in clause 52.2.

Clause 50.9 states that assessments of Defined Cost are treated as correct if not corrected within 13 weeks of the date that the *Contractor* notifies the *Project Manager* when a part of Defined Cost has been finalised. This means that only the last financial quarter would be subject to any adjustment. It also means that the *Project Manager* needs to undertake regular audits rather than at the end of a project. There are three exceptions to this: a discount, rebate or tax that can be recovered. This takes account of year-end and volume discounts that may be realised later.

This provision should address the problem you described.

225. What are the *Contractor's* share payment timings? (clause 53.3)

QUESTION

We are using ECC option C on a project. Can you explain the mechanism for payment when a project is in a 'pain' situation. When do any pain payments to the *Contractor* start, for example?

ANSWER

The timescales for making this payment are set out in clauses 54.3 and 54.4. A preliminary assessment is made when Completion is achieved, using estimates for the final Price for Work Done to Date and the total of the Prices (i.e. the 'target'). The final assessment is made using the final figures. That is included with the final amount due, which is determined when the *Supervisor* issues the Defects Certificate or the *Project Manager* issues a termination certificate (whichever comes first, see clause 50.1). These dates are the same whether the Parties (not just the *Contractor*) are in 'pain' or 'gain'.

The method of calculation is set out in clauses 54.1 and 54.2.

226. The purpose of Z clauses (Option Z)

QUESTION

I hear lots of horror stories when it comes to Z clauses and recognise that a great deal of damage can be done through their use. Can you explain the original intent of the Z clauses in the NEC? If I know this, I might be able to better defend not using them in my own organisation.

ANSWER

Option Z is intended for when the *Client* wishes to include *additional conditions of contract*.

The ECC User Guides give a warning that Z clauses should only be used when necessary, such as changes required in a particular country.

We find that many Z clauses are used to introduce legislation. This is unnecessary, as clause 12.2 requires the *Contractor* to comply with the law. Also, many Z clauses introduce constraints – these should be included within the Scope. So, again, this is unnecessary.

The index at the back of each contract within the NEC suite provides details on how the clauses are interlinked. This web/flow of clauses can often be severed by Z clauses – so, if Z clauses are included it is always a good idea to check whether this interrelationship has been impaired.

227. Deletion of clause 63.10 (Option Z)

QUESTION

We are a *Contractor* involved in an ECC project. There are several Z clauses on the project. One clause that has been deleted is clause 63.10, which explains how inconsistencies in the Scope are dealt with.

Where do we stand on this if there are ambiguities or inconsistencies in the Scope prepared by the *Client* or, indeed, in our own Scope?

ANSWER

This is where Z clauses can cause more problems than they try to solve! Clause 63.10 should, when left intact, work both ways. Ambiguities and inconsistencies are found most favourable to the Party that did not draft them, so inconsistencies and ambiguities in the Scope prepared by the *Client* are found most favourable to the *Contractor*, and vice versa. If this clause is deleted, then the contract does not resolve the issue. If it then ended up in dispute resolution, it is likely that the principle in clause 63.10 would apply, as this is a recognised principle in common law – nothing is certain though.

228. Reducing the time bar (Option Z)

QUESTION

We are involved in a contract in which the 8 weeks has been deleted in clause 61.3 and replaced with 2 days. The deemed acceptance provisions against the *Project Manager* have also been deleted under clause 62.6. Would these be enforceable?

ANSWER

Quite simply, if that is what has been signed, then that is what will apply. You can probably guess how the contract is going to be run and how the *Project Manager* may behave if those are the amendments that have been made!

229. Deleting physical conditions and weather (Option Z)

QUESTION

We are involved in an ECC Option C contract in which clauses 60.1(12) and 60.1(13) have been deleted. Does this literally mean that there will never be a compensation event for physical conditions and weather?

ANSWER

Yes, it does.

However, there are two instances when it perhaps does not accord with the perceptions of the drafter. One is that, for every compensation event that occurs, the *Contractor* (under clause 63.8) is perfectly entitled to include risk for both time and cost. Is this really what the drafter of the Z clause intended – compensation events including more time and being more costly?

The second is that the cost will be incurred under Option C anyway, dependent on the share ranges. In our opinion, the deletion of these clauses is short-sighted.

230. Complying with the law (Option Z)

QUESTION

I have previously been advised by lawyers that a Z clause requiring the *Contractor* to comply with the law is necessary. Have I been wrongly advised?

ANSWER

There is nothing that says you cannot do this, but it is hard to see the justification for it or how the Parties might suffer if such a clause were not present.

As an example, say there is a Z clause stating that the *Contractor's* staff shall not exceed highway speed limits. If an operative of the *Contractor* speeds on the way to work, what are you going to do within the contract, given that the *Contractor* has breached both the contract and the law?

The only justification for requiring compliance with the law as a contract term is that a breach of the law would also be a breach of contract. If some aspect of the law is so critical that you want to recover damages for the consequences of a failure to comply, that needs to be spelt out in a Z clause, so people understand what is intended.

231. Stepping-down Z clauses (Option Z)

QUESTION

Are Z clauses the correct location for 'step-down' clauses from a main contract (i.e. an Institution of Chemical Engineers (IChemE) contract) to a subcontract (i.e. the PSC)?

ANSWER

Yes, that would be the correct location. However, for the example given, you would need to decide carefully which process-related clauses from the main contract would be appropriate for the professional services appointment.

You can potentially mix and match NEC contracts with others, but you will need to take great care to ensure that they fit together as intended. The simplest approach would be to use, say, the ECC as the head contract (which of course can be used for all process contracts) or to use an IChemE form for the subcontracts.

Even after your best efforts when using a mix of contracts, you may still end up with one part of the supply chain using NEC language and processes (early warnings, compensation events, accepted programme) and another using entirely different language covering entirely different matters.

232. Mutual trust and co-operation (Option Z)

QUESTION

Surely some Z clauses are not compliant with ECC clause 10.2 requiring the Parties to act in a spirit of mutual trust and co-operation?

ANSWER

Some badly drafted or inappropriate Z clauses may indeed compromise the mutual trust and co-operation requirement, such that an inconsistency or ambiguity might arise. However, not all Z clauses will introduce inconsistency or ambiguity, nor will they compromise clause 10.2. Z clauses should only be included by a *Client* if they are considered absolutely necessary, and then they should be well thought out and well drafted to avoid compromising standard NEC terms.

233. Spirit of the contract (Option Z)

QUESTION

What is the status of Z clauses that break the spirit of the contract? Case law is against a Party that badly amends a contract. From the *Contractor's* point of view, is it a risk worth signing up to?

ANSWER

There is a point at which the spirit of the contract may indeed be broken by Z clauses. We would question why any *Client* would choose a standard form of contract and then decimate it. Tenderers should carefully consider whether Z clauses are acceptable to them, and act accordingly. Ambiguities and inconsistencies may well be resolved by those who created them – it is far better to sort such matters out before bidding rather than after.

234. Amending a core clause (Option Z)

QUESTION

Would it be correct to say that an amendment of a core clause is a Z clause?

ANSWER

That would be correct. Any amendment, addition or deletion of any core clause should be contained in a Z clause.

235. Replacing sentences (Option Z)

QUESTION

Instead of Z clauses, can you remove and replace core clause sentences?

ANSWER

You can indeed, but what would you call this and how would you make sure that such changes were properly incorporated into the contract? By pointing to the Z clauses in Contract Data part one and making sure that all such clauses sit clearly under the heading 'Option Z: *additional conditions of contract*', there should then be no doubt of their inclusion.

236. Protecting clients (Option Z)

QUESTION

For UK government contracts, *Project Managers* are prone to adding Z clauses that are too protective of the *Client*. How do we balance such a tendency?

ANSWER

Technically, the *Project Manager* does not have a role in the contract drafting as the role is not created until the Contract Date. The best way to protect any *Client* is to ensure that the contract has a balanced risk allocation, with each Party retaining the risks it can manage and allow for. A *Client* that tends to add Z clauses that result in an unbalanced risk allocation should be encouraged to focus instead on what is best for the project, which might help to balance such tendencies. The supply chain has a voice, and this should be used at the latest at the tender stage to offer an alternative view when risk allocation appears to be unbalanced by Z clauses.

237. Using other main Option clauses (Option Z)

QUESTION

Can you write a Z clause to amalgamate certain provisions from one NEC contract main Option into another – for example, using progress payment provisions from ECC Option B in ECC Option A? The reason would be to change the usual Option A requirement of requiring completion of a whole activity prior to payment.

ANSWER

Yes, it would be possible to do this, although we advise against it. You would, though, need to be clear what benefit this would bring, as presumably the intention would be to pay part-completed lump sums, and you would have to be sure how this would objectively work in practice. Instead, why not break down the lump sums into smaller activities in the *activity schedule*?

238. Valid Z clauses (Option Z)

QUESTION

We regularly use Z clauses to deal with issues such as professional indemnity insurance requirements, assignment and collateral warranties. Are these valid uses for Z clauses or should we remove them?

ANSWER

It is important to determine what the perceived inadequacy is within, say, the ECC to decide whether indeed a Z clause is required. On the face of it, and if they were not already provided for, then these are the sorts of matters that could well be appropriate Z clauses. That said, professional indemnity insurance is a part of Option X15 when used in the ECC; assignment is dealt with in clause 28; and as for collateral warranties, are these absolutely necessary and could the provisions of Option Y(UK)3 be used instead, saving on drafting fees?

239. Y(UK)2 and compliance with legislation (Option Y(UK)2)

QUESTION

Option Y(UK)2 states: 'The date on which a payment becomes due is seven days after the assessment date.' The Housing Grants, Construction and Regeneration Act 1996 (HGCRA) states: 'The contract requires the payer or a specified person to give the payee a notice complying with section 110A(2) not later than five days after the payment due date.'

When should the *Project Manager* certify the payment? In accordance with Option Y(UK)2, or the actual HGCRA?

Do we say that 'seven days after the assessment date' is the same as 'not later than five days after the payment due date'?

I tend to say to people, if you comply with Option Y(UK)2 then you are correct in terms of the HGCR Act, but shouldn't people assume that the two should match?

ANSWER

The ECC's provisions in section 5 of the core clauses and Y(UK)2 comply with HGCRA and its equivalent in Northern Ireland. HGCRA does not say what you suggest; you have quoted from the Act out of context (s110B(1)(a)). Section 110A(1)(a) is relevant here and requires a payment notice to be given within 5 days after the due date. The due date is created in Y(UK)2.2 and it is 7 days after the assessment date. The *Project Manager* is required to certify the amount due within a week of the assessment date (clause 51.1) so this is done on or before the due date, therefore complying with the 5-day rule from HGCRA.

240. Issuing a pay less notice (Option Y(UK)2)

QUESTION

Can the *Project Manager* issue a pay less notice under Y(UK)2?

ANSWER

Y2.3 specifically states that it is for the Parties, not the *Project Manager*, to give notice of an intention to pay less than the notified sum. The notified sum is assessed by the *Project Manager*.

241. Changing the *access dates* (Option X22)

QUESTION

Does the ECC permit the changing of the Access Dates?

ANSWER

Prior to modifications to Option X22, there was only the concept of *access dates*, stated in Contract Data part one by the *Client*, and these could not be changed, except by agreement of the Parties under clause 12.3.

The only express provision for changing this comes in Option X22 that introduces a new defined term called Access Dates, which start out as the same *access dates* as before. During negotiations in Stage One, it may be that the *Contractor* needs a change to the Access Dates, perhaps to have them changed to being earlier. The *Contractor* submits its proposals at the end of Stage One showing any revisions to the Access Dates. In turn, the *Project Manager* can issue a notice to proceed to Stage Two when the *Project Manager* and the *Contractor* have agreed any changes to the Access Dates.

242. Changing Site Information (Option X22)

QUESTION

Does the ECC permit the changing of the Site Information?

ANSWER

Prior to modifications to Option X22, there was only the concept of one issued Site Information, stated in Contract Data part one by the *Client*, and this could not be changed, except by agreement of the Parties under clause 12.3 (which we have never experienced).

The only express provision for changing this comes in Option X22. During negotiations in Stage One, it may be that the *Contractor* (or any other contractor engaged by the *Client*) needs to carry out some further site investigation, perhaps because there is significant re-design needed and there is not enough current Site Information available. Clause X22.3(7) allows the *Project Manager* and *Contractor* to agree to change the Site Information during Stage One and, if that happens, the *Project Manager* changes the Site Information as agreed.

243. *Budget incentive* in Option X22 (Option X22)

QUESTION

If we intend to incorporate X22 into our ECC contract, must we have put a *budget incentive* in the Contract Data?

ANSWER

No, it is entirely up to the *Client* when completing the X22 provisions in the Contract Data. Include this if you believe it will help to drive the right behaviours and help to create better outcomes.

244. Pricing whole life quotations (Option X21)

QUESTION

Do we use Defined Cost and the Fee in our quotation for an Option X21 proposal?

ANSWER

You may do, it is up to you, but such proposals are not compensation events and there is no explicit reference in Option X21 to using Defined Cost and Fee in your calculations. This is because you will need to factor in the cost reduction to the *Client* of the asset over its whole life and decide how much of the saving you would like to offer to the *Client* in the form of proposed changes to the Prices. Defined Cost may well help determine the difference in capital costs initially, so it is a judgement.

245. Delay damages after termination (Option X7)

QUESTION

Let's say there is termination due to the *Contractor's* fault and there is no argument about this. Assume we engage another contractor to finish the job. Do we retain delay damages until the other contractor has finished the job?

ANSWER

No, clause X7.1 is clear that delay damages can only run until the earlier of three scenarios, one of which is the date on which the *Project Manager* issues a termination certificate. If, therefore, there is a termination certificate issued before the Completion Date, for example, there can be no delay damages retained.

246. Climate change (Option X29)

QUESTION

Do I have to use Option X29 for climate change or can I use the Key Performance Indicators provisions in Option X12 or X20?

ANSWER

Option X29 is probably better suited and written more specifically for climate change, including for things like a Climate Change Plan, but in essence you could include similar targets in Option X12 or X20.

247. Climate change Performance Table (Option X29)

QUESTION

Do I have to include adjustments to payment if the measured performance is either higher or lower than its target?

ANSWER

No, in fact you do not have to put any monetary values in at all. If you consider there is a financial advantage or disadvantage to the *Client* or the *Contractor* beating or not beating the target, you may include such in the Performance Table, but it is not compulsory.

248. *Contractor's* proposals for climate change (Option X29)

QUESTION

If I don't wish to include targets and a *performance table* in respect of climate change, can I still incorporate Option X29 and hope the *Contractor* suggests some good proposals?

ANSWER

Yes, clause X29.11 is still operable and allows the *Contractor* to propose changes to the Scope in order to reduce the impact of the creation, operation, maintenance or demolition of the *works* on climate change.

FURTHER THOUGHTS AND TOP TIPS

- Be clear on the project objectives and risks. Your selection of main and secondary Options should be focused on these.
- If using Option C, D, E or F, ensure that a proactive open-book audit is undertaken at project start and subsequently. This will ensure a clear understanding of cost and avoid differences of opinion later.
- Ensure that the data for the chosen Options are properly completed in the Contract Data. If not, we have an incomplete contract!
- Carefully consider the need for Z clauses. If they must be included, check their impact by reviewing the index (at the back of each NEC contract).
- Carefully consider the optional statements and include in the relevant section.

emerald PUBLISHING ice Publishing

Robert Alan Gerrard and Patrick Waterhouse
ISBN 978-1-83708-959-8
https://doi.org/10.1108/978-1-83708-956-720252011

Chapter 11
Schedule of Cost Components (SCC)

249. Time working within the Working Areas (SCC item 1)

QUESTION

With reference to the Short Schedule of Cost Components (SSCC) in the Engineering and Construction Contract (ECC), and people who are directly employed by the *Contractor*, is it envisaged that time spent working away from the Working Areas would be classed as Defined Cost? If so, what would be considered as the proportion of time working within the Working Areas necessary to constitute the Working Areas being the 'normal place of working'?

ANSWER

If a person's normal place of working is within the Working Areas, then their costs will be included in Defined Cost according to the total time spent doing work in the contract. So, if the Site Manager goes to head office once a month for a meeting, the Defined Cost for that day is included, as it would be if some work was done from home. The ECC also allows the *Contractor* to identify people in the Contract Data who are employed by the *Contractor* but whose normal place of working is elsewhere. Those people's costs will qualify as Defined Cost wherever they work provided their work is part of Providing the Works. For those other people whose normal place of working is not within the Working Areas, their costs would be classed as Defined Cost only when they were working within the Working Areas. However, it must be remembered that the normal place of working of Site personnel will change regularly. For example, if Completion is achieved and the Site Manager moves to another site, from then onwards that person's normal place of working will not be the Working Areas.

Other people may have no 'normal place of working'. For example, a quantity surveyor who looks after four contracts may visit each site for a day a week and then spend a day at head office. In that case, the surveyor has no normal place of working at that time. If that person is on your project, 1 day of the appropriate People Rate will be recoverable, as that relates to the time spent within the Working Areas.

It is therefore simply not possible to make clear rules on this, and it is going to come down to an objective view as to whether somebody has a normal place of working and, if they do, where that is.

250. How is cost defined in Option C of the Term Service Contract (TSC)? (SCC)

QUESTION

I am used to the ECC but am about to use the TSC on a large maintenance contract for the first time. How is cost defined in the TSC? Is it like the ECC?

ANSWER

In the TSC, there is a detailed SCC for use under main Options C and E. So, yes, this operates in the same manner as the ECC.

The TSC has Service Areas (clause 11.2(15)) and the ability to add to these under clause 16.4 if required. This is like Working Areas under the ECC (clause 11.2(20)).

There is also an SSCC for use under main Option A.

251. Can we recover congestion charges? (SCC item 13(a))

QUESTION

We are currently working on an ECC Option C contract. Can you advise if the *Project Manager* is correct in its assessment of disallowing all costs in relation to London congestion charges (a type of vehicle tolling)? The *Project Manager* views these are not being incurred in order to Provide the Works.

Can these costs be claimed for under SCC people item 13(a) – payments made in relation to people for travel?

ANSWER

'Provide the Works' is a defined term (clause 11.2(15)). It includes 'all incidental … actions which the contract requires'. An 'incidental action' that the contract requires is for the people to travel to work, and that will include incurring congestion charges if they travel into or through the congestion zone.

This cost will therefore fall within SCC item 13(a), providing it accords with their employment contract. Equally, it could fall within item 13(i) or 13(n). Whichever it falls within, it is incurred to Provide the Works, as is defined in the contract.

252. Meeting the requirements of the law (SCC item 13)

QUESTION

We have a client that is of the opinion that the Construction Industry Training Board (CITB) levy (in the UK) we must pay regarding our people is not an acceptable part of the SCC and should be included within the Fee. Can you advise please?

ANSWER

We assume this query relates to the ECC. As far as we are aware, the law requires you to pay the CITB levy. We believe it is governed by the Industrial Training Act 1982 and subsequent Levy Orders issued by the Government under that Act, but you will need to get legal advice to confirm that. In that case, this is clearly a payment made in relation to people for contributions, levies or taxes imposed by law, and is therefore payable under SCC item 13(i).

253. Using estimated staff costs (SCC item 13)

QUESTION

We have recently started a contract using ECC Option C and have an issue regarding payment of people costs. The *Contractor* is submitting its direct staff costs for the agent, quantity surveyor and so on, on a pro rata allowance per week basis, not actual staff cost. It is justifying this by stating that the quantity surveyor, for example, is allocated 2 days a week for the duration of the project, and these are the costs we are getting submitted in the monthly applications for payment.

We believe this to be an incorrect method of valuing staff costs because the *Contractor* is not taking staff time from, say, timesheets of actual time spent in the Working Areas. On this project, the *Contractor* is charging us a standard allowance per week regardless of the number of days actually spent. Are we wrong in this?

ANSWER

The term used in the contract is Defined Cost, and for Option C the definition is set out in clause 11.2(24). The last part of this requires that you use the SCC to value such cost. SCC item 1 sets out what people are paid for and when. As this quantity surveyor is not based in the Working Areas, it is the second or third bullets that apply in their case. The ECC allows the *Contractor* to identify people in the Contract Data who are employed by the *Contractor* but whose normal place of working is elsewhere. Those people's costs will qualify as

Defined Cost wherever they work provided their work is part of Providing the Works. If the quantity surveyor is not named in the Contract Data, then their costs, where working elsewhere, are assumed to be within the Fee – see the first sentence of clause 52.1. Although you make no mention of it, you will need to make sure that the costs are paid on the amounts paid as set out in SCC items 11, 12 and 13. The *Project Manager* has the right to see and check such amounts, including looking at payroll records, under clauses 52.2 and 52.3.

254. Payment for pension shortfall (SCC item 13)

QUESTION

We have an ECC Option C contract. Would the *Client* have an obligation to pay any substantial additional pension contributions to *Contractor's* people in a year (assume it's basically a one-off bonus)?

ANSWER

The simple answer is no (SCC item 13(j)). While payments made in relation to people for their pension is a valid part of Defined Cost, it is subject to clause 52.1, and it needs to be at open market prices. This could be determined in the accounts the year before and after the special one-off bonus.

255. Assessment of compensation events under Options A and B (SCC item 1)

QUESTION

As a *Project Manager*, I'm confused as to how I assess people costs under main Options A and B of the ECC. Is it based on tendered rates for people or based on the cost of employment (i.e. a payroll cost plus uplift)?

ANSWER

The people costs are based on the People Rates (SCC, item 11). These are a defined term under clause 11.2(28) and inserted by the *Contractor* in Contract Data part two at tender. These then form the assessment of people costs for a compensation event.

If a compensation event requires a person or category that is not listed in the People Rates, then these may be agreed or assessed by the *Project Manager* (clause 63.16). These are, in turn, added to the People Rates. This covers instances where the nature of the compensation event requires people who were not originally envisaged.

256. Dealing with medical expenses (SCC item 13)

QUESTION

We are currently running an ECC Option C on Site (£9 million, 30-month programme). Our *Contractor* has put through a claim for a staff member's medical expense against the contract for £11 000. It seems that the original (private) hospital healthcare bill came to roughly £13 000, and the *Contractor's* insurance company has settled to pay £2000 of the bill. The *Contractor* has looked to settle the remainder of the £11 000 bill by adding this as a contract cost, which it is claiming in relation to people for medical aid under SCC item 13(m). Is this acceptable?

We have no concerns that the cost was incurred, and that the person is a full-time staff member of the project. We also feel that if the *Contractor* had suitable healthcare insurance that covered the full cost, then we would have been happy that the premium was a contract cost. However, the excess should have been covered by the individual or the *Contractor* and not the *Client*. This has resulted in £11 000 of unexpected cost to the contract, which was not allowed for.

ANSWER

Paying this will be a judgement call, we suspect. Firstly, why did the medical insurance company only pay for a small amount of this? Was the rest considered to be too expensive or unnecessary? If that is the case, then you could say that it does not comply with clause 52.1 and should not be paid. Next, what was this for? Was it to enable the person to be cured quickly of something that was going to otherwise keep that person off work or prevent them working properly (until the (UK Government) National Health Service could eventually cure it) or was it purely cosmetic? The former is likely to be payable, the latter is not because the 'market rate' would be the National Health Service's cost, which would be nil (again, see clause 52.1). Also, was it a procedure that you would normally expect the National Health Service to carry out just as quickly and as well? In that case, again clause 52.1 would say that this was not at the 'market rate', which was nil. And finally, what does the person's contract of employment say about all of this? In our experience, it is very unusual for the employer of the person (in this case the *Contractor*) to pick up any excess on medical insurance, or any other amounts that the insurer will not pay. Most employers insist that the employee pays those amounts. If it is not an obligation for the *Contractor* to pay this, then you should not be required to pay it. After all, the *Contractor* (not the employee) must be shown to have paid this amount, and it has not done so. If all of these questions answer in the favour of the *Contractor*, then we think it should be paid by you.

257. Hand tools (SCC item 2)

QUESTION

We are using the ECC. Where do we recover the cost of things like hand tools? We are encountering several arguments with contractors about this issue and would like to resolve.

ANSWER

This sort of item falls within the definition of Equipment (clause 11.2(9)) and is recoverable under item 2 of the SCC.

258. Is security recoverable? (SCC item 1)

QUESTION

We have a contract in place in which the *Contractor* is seeking payment for the security staff under SCC item 11. Is this correct or should this be included in the Fee?

ANSWER

People carrying out such security work will fit within the provisions of the SCC for people, providing they meet the requirements of one of the two bullets in item 1 or 14 of the SCC.

259. How is a security guard paid for? (SCC item 1)

QUESTION

In an ECC Option C contract, if the *Contractor* employs a security firm to undertake security work within the Working Areas, paid for on an hourly rate subcontract basis, how is this paid for?

ANSWER

It looks in your case as if these are paid for as people under item 1 of the SCC, and most likely would probably fall under item 14.

260. What does 'royalties' mean in the SCC? (SCC item 53)

QUESTION

In the ECC, what does 'royalties' in SCC item 53 cover?

ANSWER

The term 'royalties' covers those things such as a payment to an owner for the use of property, especially patents, copyrighted works, franchises and natural resources.

261. Allowing for the SCC Working Areas overheads percentage

QUESTION

We recall that in the previous edition of the ECC Option C contract a Working Areas overheads percentage, as stated in the Contract Data, was applied. Where do we now make provision for the things that this previously covered?

ANSWER

This provision has been deleted in the NEC4 ECC, and therefore all the matters it was deemed to include are distributed across the other parts of the SCC.

262. Damage to Equipment (SCC item 8)

QUESTION

We have a query on an ECC Option C contract concerning Disallowed Cost. We are questioning repairs to a *Contractor's* camera, damaged on Site. Is this a valid Defined Cost or not? We can see the hire is allowable under SCC item 22, but can see no item whereby the repair is an allowable cost.

ANSWER

You have rightly identified this camera as Equipment. The Insurance Table in clause 83.3 requires the *Contractor* to insure loss of or damage to Equipment. The first bullet of SCC item 8 makes it clear that you deduct from cost the cost of events for which the contract requires the *Contractor* to insure, and that includes this event. You do not therefore pay for this. The fact that the damage will probably be below the excess on the insurance policy is at the *Contractor's* risk.

263. Damage to vehicles (SCC item 8)

QUESTION

We are using ECC Option C. The *Contractor* has included within Defined Cost the cost of repairing damage to vehicles that have been used by the *Contractor's* employees over the past 2–3 years. The cost of a replacement hire vehicle has also been included. Such costs have been included within Defined Cost on the basis of SCC people item 13(n), 'a vehicle'. The vehicles in question are owned and repaired by another division of the *Contractor*.

The *Project Manager* considers that the cost of damage repair and hire falls under SCC insurance item 8, first bullet, and, as such, adequate insurance provision should have been provided under clause 83.3, row 3 of the Insurance Table. The *Project Manager* has disallowed significant repair and/or hire costs because of clause 52.1, which states that any *Contractor's* costs that are not included in the Defined Cost are treated as included in the Fee. Does the cost of repairing damaged vehicles and the cost of replacement hire vehicles fall under SCC item 13(n) as Defined Cost or should it be treated as included in the Fee?

ANSWER

These vehicles are not owned by the *Contractor* but by a separate legal entity, albeit one with the same parent company. As such, the *Contractor* would be liable to that separate entity for any damage that occurs to its property. Therefore, under the third row of the Insurance Table, the *Contractor* is required to insure against such liability. The costs of these accidents could well fall within SCC item 13(n); however, the first bullet of SCC item 8 makes it clear that such costs are, nevertheless, to be deducted because they are the cost of events that the contract requires the *Contractor* to insure against. The *Project Manager* is therefore correct not to allow such costs.

264. Can we recover the costs of transport for people? (SSCC item 1)

QUESTION

We are using the ECC and have some compensation events valued using the SSCC. An item we have claimed for under the Equipment heading is for a work van used to transport people to and from the Working Areas. The *Project Manager* has struck this cost out in our quotations. The published list in the Contract Data is the Civil Engineering Contractors Association (CECA) Schedule of Dayworks. Under item 3, 'Supplementary Charges within the General Notes', it says at point 1: 'Transport provided by contractors for operatives to, from, in and around the site to be charged at the appropriate Schedule rates.' Is the *Project Manager* correct in not including the cost of the van in our quotations, or can we recover this as a cost?

ANSWER

You cannot claim for this van as Equipment, because that only covers Equipment used in the Working Areas, not transporting people to those areas (first sentence of SSCC item 2). It does not matter what the General Notes in the CECA Schedule says, because you only use the Schedule to get the rates for the Equipment that is to be paid for in accordance with the SSCC.

The cost of this, however, should have been included as part of the cost of SSCC item 1 (people), and allowed for in the People Rates.

265. Cost of people in the SSCC (SSCC item 1)

QUESTION

Under the SSCC in ECC Option A, there is a provision for 'People Rates' in item 1. It is unclear what can be included in the cost of people, and it is our view that it is made up of the cost to employ the person (SSCC item 11); that is, the salary of the person plus tax, national insurance and pension contributions, plus any taxable allowances – for example, a car allowance. We do not believe the cost of people includes the cost of such items as travel, subsistence and lodging, as we believe these are included elsewhere. Are our views in line with the intended application of the SSCC?

ANSWER

For People Rates, if they are a cost for employing the person, which they are, then they should have been included in the appropriate People Rates.

266. Which SSCC Equipment rates do we use? (SSCC item 2)

QUESTION

We are currently engaged on an ECC Option A contract. At the tender stage, in our Contract Data part two – data for SSCC, we submitted a percentage reduction for Equipment under the Civil Engineering Contractors Association (CECA) Schedule of Dayworks. We also submitted a rate under 'other Equipment' for a gang consisting of a 7-tonne lorry, a mini-excavator with rock breaker and a transporting trailer. This gang was dedicated to this particular contract, which involved minor works at dozens of different locations.

We have since submitted the 'other Equipment' gang rate in quotations for compensation events, but the *Project Manager* has assessed the compensation events at a lower value by breaking our gang down to its constituent parts and applying the CECA Schedule with our reduction. We can see where the *Project Manager* is coming from, but we contend that we submitted a gang rate in our tender, that gang was used for compensation event works and we are entitled to compensation event assessment at that rate. Could you advise which rate should take precedence?

ANSWER

The 'other Equipment' is supposed to be used only for Equipment that is not within the CECA rates, which is why the entry in the Contract Data follows the CECA entries and therefore refers to only Equipment 'other' than can be found in the CECA rates. It is for specialist items of Equipment, such as piling rigs, tunnel-boring machines and dredgers, which are not covered by the CECA rates. That can be seen when you look at SSCC items 21 and 22. Item 21 is used for all Equipment for which a rate can be found in the published list, and item 22 is only used for Equipment 'which is not in the published list stated in the Contract Data'. Therefore, the *Project Manager* is correct in what it is doing.

267. Cost of Equipment standing (SSCC item 2)

QUESTION

We are using ECC Option B and would appreciate your thoughts on the following event. On the first day of the construction project the *Contractor* had to stand down an excavator, which lasted a matter of weeks. A compensation event was notified with the Equipment priced at the rate listed in Contract Data part two to be recovered for the period of delay. It is our view that Equipment listed in Contract Data part two is meant for incidental works and is not necessarily intended for use when there may be significant delay. We have subsequently suggested that the Equipment should be priced at market rates or the equivalent hire cost, as depreciation/fuel costs would be reduced as a result of the stand down.

Secondly, if the *Contractor* had programmed to bring Equipment to the Site but did not as a result of the stand down, we feel that as the Equipment was not within the Working Areas, it cannot be charged for. Is this correct?

ANSWER

The contract is clear that you use the rate in the Contract Data (see either SSCC item 21 or 22, depending on which applies). Terms such as 'incidental works' are irrelevant because they are not used in this contract. If you thought the rate was too high (or the percentage was wrong), then the time to raise that was before you accepted the tender. Once you have a contract, what it states is what you pay.

Regarding your second point, an initial inspection of the contract would suggest that you do not pay this cost because the Equipment was not in the Working Areas. However, it is not as simple as that. The compensation is based on a forecast of the Defined Cost, not the actual Defined Cost. That is because the last sentence of clause 63.1 makes it clear when you stop using actual Defined Cost, and clause 63.1 also makes it clear that is the date when you issued the instruction that we are talking about. Once you use a forecast you do not change it based on what happened (clause 66.3). So, would a reasonable forecast be that the *Contractor* brought the Equipment to the Site? Given that it will not be paid for unless it does (and we are aware of the circulatory nature of that statement), then we suspect the answer is yes. If you were to allow for it, we think you have the right to check to make sure that it was, indeed, standing and not being used somewhere else. There is no definitive answer to this. We suspect that if the *Contractor* could show that it was not using it, most *Adjudicators* would stick to the literal meaning of the contract and not allow payment for it.

268. Using the percentage for people overheads

QUESTION

Our query relates to the ECC and Contract Data part two, specifically where Option A or B is used (although there are similar requirements elsewhere). The previous version of the ECC stated: 'Data for the Short Schedule of Cost Components' and 'The percentage for people overhead is ...%.'

We have read the NEC4 ECC and the new ECC User Guides and cannot find this inclusion any more – where has it gone?

ANSWER

This particular provision has been removed in the current ECC, and all the items this provision was deemed to cover have been distributed across other provisions of the contract.

269. The cost of internet charges (SSCC item 51)

QUESTION

Under an ECC Option A contract, can you advise how we recover the cost of internet charges in respect of compensation events?

ANSWER

This is provided as a recoverable cost in SSCC item 51. The difficulty you have will be to determine how much additional provision and use in the Working Areas arises because of a compensation event.

270. Using the percentage for design overheads

QUESTION

Where should we allow for the percentage for design overheads – which was in the previous edition of ECC Option A? Also, how does this differ from the Fee?

ANSWER

The percentage for design overheads has been removed in the NEC4 ECC and therefore the matters that this was deemed to cover should be allowed for in other parts of your tender – so you can more sensibly include for such in the appropriate hourly rates in the Contract Data; alternatively, and probably less sensibly, you could include such in the Fee.

271. Working from home (SCC item 1)

QUESTION

We have people who are now permanently working from home. How can we recover the cost of these in an ECC Option C contract?

ANSWER

The third bullet of item 1 in the SCC covers people whose normal place of working is not within the Working Areas and who are working outside of the Working Areas, such as at home. However, these people must be employed by the *Contractor*, and they cannot do manufacture and fabrication or design, as these are covered in items 6 and 7 of the SCC, respectively. They also must be listed in the Contract Data and, if the *Project Manager* agrees, additional people may be assessed as if they were listed in the Contract Data.

That said, it may be difficult to determine the actual time spent if they are working for multiple clients, so the *Contractor* can just make due allowance in the Fee when tendering.

FURTHER THOUGHTS AND TOP TIPS

- The purpose of both the SCC and SSCC is to assist in the clarity of understanding Defined Cost. Both will need careful study to aid the tender process and later contract management.

Robert Alan Gerrard and Patrick Waterhouse
ISBN 978-1-83708-959-8
https://doi.org/10.1108/978-1-83708-956-720252012

Chapter 12
Contract Data

272. What *service* is in the Professional Service Contract (PSC) Scope? (Contract Data)

QUESTION

A *Contractor* has been employed under the PSC to provide overall management of a programme of *service* to include whole life-cycle activities from competition to completion. My question is, does this include the provision of an Engineering and Construction Contract (ECC) *Supervisor* or on-Site health and safety auditor, or is this a role that the local authority would procure?

ANSWER

The answer will depend entirely on what is written in the relevant PSC Scope. If it says that the *Consultant* (the name used in the PSC) provides this *service*, then it is included. Otherwise, the answer will not be clear and will depend on an interpretation of the words used in the Scope to define precisely what the *Consultant* does.

273. Who is Mr I Judge? (Contract Data)

QUESTION

We are about to use the ECC for the first time on a new school project. We have never completed the Contract Data previously – could you direct us to any relevant guidance?

ANSWER

Each of the contracts within the NEC suite has accompanying User Guides. The ECC User Guides contain advice on how to complete the Contract Data. This is annotated in order that you can refer to the relevant section in the User Guides themselves. Do not copy it exactly, as we have seen a real contract with ‘Mr I Judge’ as the *Adjudicator*. This was meant to be a joke!

274. How do you describe the Affected Property? (Contract Data)

QUESTION

We are preparing tender documents using the Term Service Contract (TSC). When an Affected Property is made up of 10–15 buildings (which form an establishment), what can we refer to these as?

ANSWER

The first thing you need to decide is exactly what the Affected Property is. Is it 10 or 15, or any number in between? Once you know that, you can describe it in any way that is sensible and informs the *Contractor* exactly what buildings it is required to provide services to. So, you can show them on a drawing and refer to the drawing number in the Contract Data. Or list them in the Contract Data. Or say 'all the buildings at XYZ's site on the High Street, Anywhere' in the Contract Data. How does not matter, if it is clear.

What many people do is produce a table of buildings describing each one and explaining what is required in each, and what the constraints are for each (such as access). You can then put that table in the Scope, and the entry in the Contract Data can say something like 'The Affected Property is shown in Section 3 of the Scope'. All of this, of course, will depend on exactly what maintenance or operations or facilities management you want the *Contractor* to carry out in these buildings.

275. Incomplete Subcontract Data (Subcontract Data)

QUESTION

We are a subcontractor under the ECS. The Subcontract Data (first bullet under 'General') states that retention and delay damages will apply. However, this has not been completed – how would this be interpreted?

ANSWER

The ECS User Guide warns that failure to complete fully will result in an incomplete contract – this is what you have here. There is no real solution here other than to agree to change the contract by agreement of the Parties (this would be a suitably authorised individual in your company and within the main *Contractor's* organisation) – this needs to be recorded in writing and signed by the two Parties (clause 12.3).

276. Do we need to name the *Project Manager*? (Contract Data)

QUESTION

I note in the ECC User Guides on completing the Contract Data that individuals are named. Should this be the case, or can we name a department? Also, what happens if the named individuals move on or need to delegate?

ANSWER

Yes, you may name the individuals or the organisation they work for – the ECC is all about having absolute clarity with communication and clear lines of authority. If those individuals are replaced, then clause 14.4 requires that the *Client* notifies the *Contractor* of the name of the replacement.

Any delegation of duty by the *Supervisor* or *Project Manager* should be notified in accordance with clause 14.2. This should be clearly articulated – it is recommended that this is on a clause-by-clause basis.

Note that both clauses use the term 'notify'. In accordance with clause 13.7, this means that the communication should be undertaken separately from other communications. Again, it is about clarity.

277. Scope versus the *works* (Contract Data)

QUESTION

I am completing the Contract Data for an ECC project – can you advise on the difference between the *works* (second bullet) and the Scope?

ANSWER

The *works* should be a clear and succinct description of the works required – for example, the design and construction of a new oncology facility at the Newtown site.

The detail is contained in the Scope. The Scope is a defined term under the ECC (i.e. it has a contractual meaning). This is defined in clause 11.2(16) – it specifies and describes the *works*. It also states any constraints on how the *Contractor* Provides the Works.

278. Best practice in drafting the Contract Data (Contract Data)

QUESTION

We want to ensure that we comply with best practice in drafting the Contract Data for the ECC as a *Client* team. What do you suggest to ensure we incorporate best practice and any relevant guidance?

ANSWER

The Scope is one of the most important documents referenced from the Contract Data. There is very helpful guidance in the ECC User Guides. It is recommended that the structure and drafting guidance is followed. This should ensure that the right terminology is used and that this accords with ECC drafting principles.

Good practice in terms of the Contract Data would typically include initially striking through items (i.e. secondary Options and optional statements) that you do not intend to use. When circulating to a wider team, this will highlight which options are not being used and may stimulate debate. Once a final version has been agreed, these should be deleted to provide a 'clean' copy for the Parties.

279. The *period for reply* (Contract Data)

QUESTION

Can you explain the *period for reply* – when is this applicable? As far as I am aware, some of the clauses in the ECC have timescales – does the *period for reply* take precedence?

ANSWER

Clause 13.3 is perhaps the best starting point to explain the *period for reply*. Clause 13.3 states that, unless a timescale is stated in the contract, the *Supervisor*, *Contractor* and *Project Manager* all respond within the *period for reply*. Typically, this is 2 weeks – sometimes this is less on projects that perhaps are fast track.

It is worth noting that differing *periods for reply* may be introduced – for example, for design acceptance, the Contract Data may allow the *Project Manager* more time (i.e. 3 weeks). A note of caution – too long a period could make the project very difficult to deliver, so a pragmatic approach is required.

280. Responding quicker than the *period for reply* (Contract Data)

QUESTION

What happens if the *Contractor* needs a decision quicker than the *period for reply*? Is the ECC *Project Manager* obliged only to respond within the *period for reply*?

ANSWER

The maximum time period for the *Project Manager* to respond (unless a timescale is defined in the contract) is the *period for reply*. In a spirit of mutual trust and co-operation, this should be seen as the maximum.

If the *Contractor* requires a quicker turn around, then the correct procedure in the contract is to notify an early warning. In a spirit of mutual trust and co-operation, it is anticipated that the *Project Manager* would then make all efforts to decide more quickly.

281. Contractual significance of the Early Warning Register (Contract Data)

QUESTION

Can you explain the relevance of items that will be included in the Early Warning Register versus *Client's* liabilities in the ECC? What status do they have, and how do they affect other risk allocation?

ANSWER

Items that will be included in the Early Warning Register should be listed in the Contract Data by both the *Contractor* and *Client* at the tender stage. They *do not* change the risk allocation. These simply comprise a proactive list of matters that, once the contract is signed, will be entered into the Early Warning Register and mitigation plans developed.

Items that are listed as *Client's* liabilities *do* change the risk allocation. If included and then they occur, these become a compensation event under clause 60.1(14) and the *Client* becomes responsible for some other costs incurred by the *Contractor*.

The Early Warning Register is a defined term (clause 11.2(8)). It merely needs to include a description of the matter and the mitigation plan. Many teams overcomplicate this and lose track of practical mitigation measures.

So, matters included in the Early Warning Register do not change risk allocation – additional *Client's* liabilities do.

282. The *defects date* versus the *defect correction period* (Contract Data)

QUESTION

Can you explain the difference between the *defects date* and the *defect correction period*? Also, why are there potentially several *defect correction periods* in the ECC?

ANSWER

The *defects date* is a defined number of weeks after Completion of the whole of the *works*, such as 52 weeks (note: not after *sectional* Completion).

Effectively, this is the period within which the *Supervisor* would still need to notify any Defects found and the *Contractor* is obliged to rectify within the *defect correction period*.

The *defect correction period* is therefore a defined period of time within which the Defect should be rectified. The *Contractor* cannot stockpile these and rectify in week 51 of 52 in order to gain retention release – they need to be dealt with on a case-by-case basis and within defined timescales.

283. Does the *defects date* change? (Contract Data)

QUESTION

We are using an ECC Option A contract on a develop-and-construct project. If an instruction for additional works is given after the Completion Date, and the instruction changes the Scope and is a compensation event, what effect does this have on the *defects date*?

ANSWER

Assuming Completion of the whole of the *works* has been achieved, the instruction will have no effect on the *defects date*. This is because the *defects date* is defined as being 'x' weeks after Completion of the whole of the *works* in Contract Data part one. Of course, 'x' is the number of weeks inserted in the Contract Data in your particular contract. Completion is not affected, as it has already been certified. The assessment of the compensation event will have to reflect the additional Defined Cost to the *Contractor* of carrying out and managing isolated works after its resources have left the Site.

284. Optional statements – what are they? (Contract Data)

QUESTION

What are optional statements in Contract Data part one of the ECC and when should we consider these?

ANSWER

The optional statements are exactly that – you can decide to include them or not. These decisions should be in keeping with your contract strategy and cover 'what if' scenarios. For example, what if the *Contractor* achieves Completion prior to the Completion Date – are we willing to take over? If you are not, the optional statement should be included; if you are, then there is no need to include it.

Other optional statements relate to payment timescales, additional *Client's* liabilities, Key Dates and insurances and, among others, the secondary Options.

These need to be carefully considered and should reflect project risks and key drivers/objectives.

285. Identifying *key persons* (Contract Data)

QUESTION

Do we have to let the *Contractor* define the *key persons* or can we have a say in who they should be?

ANSWER

Although Contract Data part two is completed by the *Contractor*, its contents are established prior to tender by the *Client*. You could, for example, identify who the key roles are, for example the site manager, planner, commercial manager, health and safety representative, and design manager.

You could also go further and identify their qualifications and experience. This may be very important on some projects.

This has the advantage of identifying what you see as the key attributes of a successful team on your project and a level of consistency across the competing tenderers.

286. Modelling the Contract Data part two rates and percentages (Contract Data)

QUESTION

How should we evaluate the financial submission part of the tender from the *Contractor* for a contract under Option C of the ECC? Should we just review the tendered total of the Prices, or should we develop a cost model to assess other rates and percentages?

ANSWER

You should most certainly evaluate the rates and percentages in a cost model as you suggest. This provides sensitivity analysis and a rounded picture as to the most economical *Contractor*.

For example, *Contractor* A may submit a tendered total of the Prices at £1 million but its rates and percentages are treble those of *Contractor* B whose tender comes in at £1.01 million. If we estimate £300 000 of compensation events, then *Contractor* B would be the most efficient. The ECC User Guides provide a sample cost model that could be developed for any main Option.

In order that all the tenderers are aware of the cost model, this would need to be defined in the tender documents.

287. Optional statements (Contract Data)

QUESTION

Can you explain the optional statement in Contract Data part two of the ECC – when would the *Contractor* provide the Scope for its design? Also, what happens if this is inconsistent with the *Client's* Scope?

ANSWER

The *Contractor* may be required to provide the Scope for its design if tendering procedures require this, most likely in a proposed design and construct contract. This would be provided and included in the contract before signature.

If the Scope prepared by the *Contractor* for its design is of a lower quality than the Scope prepared by the *Client*, then, under clause 17.1, the *Contractor* and *Project Manager* are required to notify each other as soon as they become aware; the *Project Manager* will then decide the outcome. Assuming its instruction is for the *Contractor* to change its Scope to comply with the *Client's*, then this will not be a compensation event (clause 60.1(1), second bullet).

288. Completing the Contract Data (Contract Data)

QUESTION

We are using the ECC and are unsure about how the Schedule of Cost Components (SCC) and Short Schedule of Cost Components (SSCC) are implemented in the contract. For example, the data for the SCC and SSCC appear to be added at the end of Contract Data part two by the *Contractor*.

My understanding is that these data relate to elements of the *works* outside the Working Areas (such as design and fabrication) and for, say, Equipment that is purchased. However, if our understanding is correct, we would also expect to see some words in Contract Data part two along the lines of 'The SCC is in the document …' and 'The SSCC is in the document …'. There would then be a list of rates for various people and Equipment and so on working on the Site; however, this does not appear to be the case. In short, we are unsure what the *Contractor* puts where – for example, a labourer costs £20 per hour or an excavator costs £35 per hour.

ANSWER

Either the SCC or the SSCC form part of the *conditions of contract* where main Option A, B, C, D or E is used. The main Option included in the contract will determine which schedule you use and what you use it for. The SCC covers all Defined Cost and sets out what will be paid and how it will be paid in main Option C, D or E. The SSCC covers Defined Cost in contracts formed using main Option A or B; this is just for assessing compensation events. The information set out in the Contract Data provides the information that is needed to make the SCC or SSCC work and is not just about work outside the Working Areas. People are paid for at what they cost in the SCC and at the People Rates stated in the SSCC. Equipment is also paid at cost in the SCC (item 2), whereas in the SSCC it is paid for at the rates in the schedule referred to in Contract Data part two, with the percentage adjustment also set out there (SSCC item 2).

289. The *Supervisor* (Contract Data)

QUESTION

Does the *Supervisor* in the ECC have to be a separate person or organisation from the *Project Manager*?

ANSWER

No, in fact the *Supervisor* could be the same person as the *Project Manager*. It depends on the competencies of the person(s), the complexity of the project and other factors. There is no requirement to have different people or organisations in these roles, but care must be taken to make sure they have the time and competencies to fulfil the roles properly.

290. No *Supervisor* named (Contract Data)

QUESTION

What happens if a *Client* doesn't appoint a *Supervisor* in the ECC?

ANSWER

The *Supervisor* should be named in the Contract Data, either a company name or a person. The *Supervisor* may be replaced during the contract, but only by the *Client* (clause 14.4). If one is not named or a replacement cannot be found, the sensible thing would be for the *Project Manager* to be temporarily named as *Supervisor* until one is found. This is providing the *Project Manager* has the right competencies to act in this capacity.

FURTHER THOUGHTS AND TOP TIPS

- Ensure that the Contract Data is completed in full. If incomplete, you have an incomplete contract!
- Carefully select the main and secondary Options and optional statements and ensure that they fit with the project risks and objectives.
- Develop a cost model that evaluates the Contract Data part two rates and percentages. Ensure that this is clearly defined in the tender documents.
- Ensure that the Scope is appropriately drafted – look at the appropriate User Guides that provide a clear structure and tips on drafting.

Printed and bound by CPI Group (UK) Ltd, Croydon, CR0 4YY

14/10/2025

14752768-0001